Engineering Education Trends in the Digital Era

Şeyda SerdarAsan
Istanbul Technical University, Turkey

Erkan Işıklı
Istanbul Technical University, Turkey

A volume in the Advances in Higher Education and Professional Development (AHEPD) Book Series

IGI Global
DISSEMINATOR OF KNOWLEDGE

Published in the United States of America by
IGI Global
Engineering Science Reference (an imprint of IGI Global)
701 E. Chocolate Avenue
Hershey PA, USA 17033
Tel: 717-533-8845
Fax: 717-533-8661
E-mail: cust@igi-global.com
Web site: http://www.igi-global.com

Library of Congress Cataloging-in-Publication Data

Names: Serdar Asan, Şeyda, 1977- editor. | Işıklı, Erkan, 1982- editor.
Title: Engineering education trends in the digital era / Şeyda SerdarAsan and Erkan Işıklı, editors.
Description: Hershey, PA : Engineering Science Reference, 2020. | Includes bibliographical references and index. | Summary: "This book addresses opportunities, challenges, and barriers encountered in engineering education in the digital era"-- Provided by publisher.
Identifiers: LCCN 2019042483 (print) | LCCN 2019042484 (ebook) | ISBN 9781799825623 (hardcover) | ISBN 9781799825630 (paperback) | ISBN 9781799825647 (ebook)
Subjects: LCSH: Engineering--Study and teaching (Higher)--Case studies. | Education, Higher--Effect of technological innovations on--Case studies.
Classification: LCC T65 .E593 2020 (print) | LCC T65 (ebook) | DDC 620.0071/1--dc23
LC record available at https://lccn.loc.gov/2019042483
LC ebook record available at https://lccn.loc.gov/2019042484

This book is published in the IGI Global book series Advances in Higher Education and Professional Development (AHEPD) (ISSN: 2327-6983; eISSN: 2327-6991)

British Cataloguing in Publication Data
A Cataloguing in Publication record for this book is available from the British Library.

All work contributed to this book is new, previously-unpublished material.
The views expressed in this book are those of the authors, but not necessarily of the publisher.

For electronic access to this publication, please contact: eresources@igi-global.com.

Advances in Higher Education and Professional Development (AHEPD) Book Series

ISSN:2327-6983
EISSN:2327-6991

Editor-in-Chief: *Jared Keengwe* University of North Dakota, USA

MISSION

As world economies continue to shift and change in response to global financial situations, job markets have begun to demand a more highly-skilled workforce. In many industries a college degree is the minimum requirement and further educational development is expected to advance. With these current trends in mind, the **Advances in Higher Education & Professional Development (AHEPD) Book Series** provides an outlet for researchers and academics to publish their research in these areas and to distribute these works to practitioners and other researchers.

AHEPD encompasses all research dealing with higher education pedagogy, development, and curriculum design, as well as all areas of professional development, regardless of focus.

COVERAGE

- Adult Education
- Assessment in Higher Education
- Career Training
- Coaching and Mentoring
- Continuing Professional Development
- Governance in Higher Education
- Higher Education Policy
- Pedagogy of Teaching Higher Education
- Vocational Education

IGI Global is currently accepting manuscripts for publication within this series. To submit a proposal for a volume in this series, please contact our Acquisition Editors at Acquisitions@igi-global.com or visit: http://www.igi-global.com/publish/.

The Advances in Higher Education and Professional Development (AHEPD) Book Series (ISSN 2327-6983) is published by IGI Global, 701 E. Chocolate Avenue, Hershey, PA 17033-1240, USA, www.igi-global.com. This series is composed of titles available for purchase individually; each title is edited to be contextually exclusive from any other title within the series. For pricing and ordering information please visit http://www.igi-global.com/book-series/advances-higher-education-professional-development/73681. Postmaster: Send all address changes to above address.

Titles in this Series

For a list of additional titles in this series, please visit:
http://www.igi-global.com/book-series/advances-higher-education-professional-development/73681

Enhancing Learning Design for Innovative Teaching in Higher Education
Sophia Palahicky (Royal Roads University, Canada)
Information Science Reference • © 2020 • 300pp • H/C (ISBN: 9781799829430) • US $195.00

Teacher Training for English-Medium Instruction in Higher Education
Maria del Mar Sánchez-Pérez (University of Almeria, Spain)
Information Science Reference • © 2020 • 447pp • H/C (ISBN: 9781799823186) • US $200.00

Handbook of Research on Adult Learning in Higher Education
Mabel C.P.O. Okojie (Mississippi State University, USA) and Tinukwa C. Boulder (University of Pittsburgh, USA)
Information Science Reference • © 2020 • 756pp • H/C (ISBN: 9781799813064) • US $245.00

Preparing Students for Community-Engaged Scholarship in Higher Education
Aaron Samuel Zimmerman (Texas Tech University, USA)
Information Science Reference • © 2020 • 465pp • H/C (ISBN: 9781799822080) • US $195.00

Management Training Programs in Higher Education for the Fourth Industrial Revolution Emerging Research and Opportunities
Edgar Oliver Cardoso Espinosa (Instituto Politécnico Nacional, Mexico)
Information Science Reference • © 2020 • 127pp • H/C (ISBN: 9781799818755) • US $165.00

Assessment, Testing, and Measurement Strategies in Global Higher Education
Elena Aurel Railean (American University of Moldova, Moldova)
Information Science Reference • © 2020 • 360pp • H/C (ISBN: 9781799823148) • US $185.00

For an entire list of titles in this series, please visit:
http://www.igi-global.com/book-series/advances-higher-education-professional-development/73681

IGI Global
DISSEMINATOR OF KNOWLEDGE

701 East Chocolate Avenue, Hershey, PA 17033, USA
Tel: 717-533-8845 x100 • Fax: 717-533-8661
E-Mail: cust@igi-global.com • www.igi-global.com

Table of Contents

Detailed Table of Contents

In this study, the stages of change in the curriculum of ITU Metallurgical and Materials Engineering Department (MME) during the ABET accreditation process are summarized and evaluated in terms of both course contents and assessment-evaluation procedures. Improvements in the curriculum design have been carried out within the framework of continuous development, which is one of the essential criteria of today, and then assessment and evaluation methods have been developed, expecting the students become more equipped in their professional life. First of all, a discussion on the characteristics of the Information Age and ABET EAC Student Outcomes (SOs) is provided, followed by the presentation of ITU MME curriculum, conveying the changes in a chronological order. Subsequently, the continuous improvement cycle of ITU MME and the measurement and assessment stages for each outcome are explained. In conclusion, the new ABET EAC SOs are given with a comparison and a final evaluation.

Higher education institutions and their engineering departments have a vital role in fulfilling the new requirements and opportunities of the information and communication technologies (ICT). Therefore, understanding the guidelines to adapt to the new ICT innovations in relation with the Fourth Industrial Revolution such as the internet of things, cloud computing, virtual reality, and artificial intelligence is vital to determine the emerging patterns in their development, delivery, implementation, and assessment. This study aims to define the new educational requirements in engineering education based on the developments of the Fourth Industrial Revolution. Conducting an in-depth analysis of the current literature revealed that although these emerging technologies have been widely used, there are some challenges being faced for their effective use in engineering education. Therefore, the authors provide some guidelines and discuss possible research directions for the use of these technologies in near future.

Higher education (HE) should focus on solving the following critical educational problems: (1) using technology and (2) fostering education by new creative learning techniques. In this chapter, the authors indirectly talk about using new technologies in education. There are many reasons that make this choice challenging such as believing in the benefits, having enough knowledge, accessing alternative technological sources, etc. To facilitate this, they introduce an online learning platform for engineering instructors in HE. Moreover, according to their experiences in the field of education and engineering, instructors in HE should improve and revise their skills and knowledge. It is important to enhance knowledge on content, technology, and pedagogy; therefore, this training platform itself focuses on improving those skills necessary for instructors in HE for maintaining an effective learning process. This chapter mainly focuses on creating a course for higher education engineering instructors and a TERR model that is proposed by the authors.

The education-training process and all activities related to it have the power to direct the future of societies. From this point of view, the process should be analyzed frequently in terms of input, output, and other process elements. Educational data mining is a multidisciplinary research area that develops methods and techniques

students' educational output are explained. Following that, the collaborative problem-solving approach and collaborative technologies are presented. Then, the collaborative problem-solving method framework and how collaborative technologies can be used with this method in the learning environment of engineering education are explained in detail. Finally, recommendations about future work are presented.

This chapter illustrates a research focused on how to effectively implement the challenge-based learning (CBL) approach in a higher education institution (HEI) in the UK. The challenge was linked to contemporary research conducted by a group of academics, which concerned how digital technologies can positively impact the local economy. The project was named 'UoL4.0 Challenge', and it proved that designing and implementing CBL educational environments can increase students' propensity to work actively and proactively. The exercise also suggested that CBL may support students in the application of their academic skills and digital capabilities to support their communities. This study presents a description of the case and a reflection on lessons learned with an aim to provide guidelines for other educators and policymakers that are interested in implementing I4.0 educational initiatives at local or national levels. It is also suggested that CBL may play a fundamental role in implementing the triple-helix model of innovation.

The VISIR+ project was an international collaboration project for the dissemination of the remote laboratory VISIR, a tool to support teaching the theory and practice of electrical and electronic circuits. The initiative was first disseminated in Europe, and Latin American countries such as Brazil followed. This chapter essentially aims to discuss the experience of the Federal University of Santa Catarina in Brazil with the VISIR+ project. Various approaches were used for the dissemination of the initiative, including free courses for high school students, teacher training, and the creation of a virtual environment to discuss and share lesson plans that used the remote laboratory VISIR on their plots. In conclusion, the experience was observed as excellent for the institution and there was no reason to put the project ideas aside. After participating in the project, it becomes a challenge to ensure its sustainability.

Engineers in a globalizing world need innovative skills as well as learning capabilities. They often need to cooperate in different teams in the supply chain of a company. Thus, they should practice before they perform like athletes and musicians do, but they cannot practice by making mistakes since it would be an expensive way to learn. Serious business games serve as a platform for the exchange of ideas, the sharing of expertise, and the alignment of objectives. Higher productivity along with employee satisfaction are their most prominent outcomes. Engineers should be able to work in teams, where behaviors can have both positive and negative effects. Serious business games can support organizations to develop more effective team behavior, influence the level of cooperation in a group, and, in turn, increase the company's profit. In this study, the authors examine a supply chain serious game called "The Fresh Connection" and discuss how it can improve the learning process at the university and further allow people to continue their learning process as an engineering professional.

Using game elements in class to support the participation of students in learning or designing games that can help educators teach certain subjects more efficiently has been a popular topic in recent years. The former is a matter of gamification, which refers to the application of game elements to other activities so that the activity becomes more engaging or interesting. On the other hand, the latter is about designing a serious game, which can be defined as a game with an explicit and carefully thought out educational purpose. In this study, focused on merging the use of gamification and serious games for a specific engineering course, the authors discuss how the curriculum for this course should be designed so that both sides of the spectrum are facilitated. An application of this union is given with a survey showing the reaction of students to the gamified curricula integrated with a serious game.

Preface

Started with the invention of the steam machine in 1712, the First Industrial Revolution turned labor intensive production into mechanical production using water and steam power and became the turning point of the current development level of both the industry and the world. As a result of cheap steel production and development of railways, inventions stemming from the use of electrical energy and mass production were introduced and the Second Industrial Revolution (The Technology Revolution) emerged thereafter. Subsequently, electronics and information technologies invaded the industry, thereby the spread of automation accelerated and level of production increased. Those were the times humanity experienced the Third Industrial Revolution, which led to dramatic changes in the skill sets and competencies of the workforce. Nowadays, we are witnessing the Digital Era with the emergence of the fourth industrial revolution (Industry 4.0) where production is based on Cyber-Physical Systems (CPS) and the Internet of Things (IoT). We are, indeed, standing on the brink of transformation into a more and more digitized world, which was essentially propelled first by automation in production systems and then by digitalization, whereby massive changes in the workforce and socio-technical advancements may be experienced once again in the near future.

The term Industry 4.0 was originated in Germany by Henning Kagermann, Wolfgang Wahlster, and Wolf-Dieter Lukas, who advocated it during a press conference at the Hannover Messe in 2011 (Kagermann et al., 2013; Schwab, 2017, p.12). Believed to modernize production systems with smart, flexible, self-organizing, dynamic, and optimizing factories (Lu, 2017) so that decision making processes are automated, Industry 4.0 essentially depends on digital technologies that can extend the overall connectivity of logistics networks and the better utilization of pooled resources (Veile et al., 2018). It is expected to radically change all aspects of our lives, thereby affecting the scientific disciplines, industries and economies in many respects (Schwab, 2017).

Companies must be capable enough to swiftly adapt to changes in the dynamics of their industries brought about by the digital era so that they can remain sustainable and gain a competitive edge (or stay ahead of the competition). This is only possible with a thorough understanding and implementation of several components such as Artificial Intelligence, Augmented/Virtual Reality, Big Data Analytics, Blockchain, Cloud Computing, Cognitive Computing, Cyber-Physical Systems (CPS), Human Machine Interface, Industrial Information Integration, Intelligent Manufacturing, Intelligent Sensors, Internet of Services (IoS), Internet of Things (IoT), Mobile Computing, Robotics, and Three-Dimensional Printing (Pfohl et al., 2014; Lu, 2017; Xu et al., 2018). Therefore, the amount of knowledge and expertise on these topics will be highly required in both the industry and the academia in the near future.

Current maturity levels of the digital transformation changes among countries and industries; however, once the adjustment to it completes, the way how people live and work will have been dramatically changed. Many jobs that have existed until recently may vanquish and many jobs may no longer need people. New jobs are also likely to appear. As discussed in the 2016 World Economic Forum, 65% of children who have just started school are expected to work in jobs that do not currently exist (WEF, 2016, p.32). Creating the future professional means not only attracting and developing new skills, but also re-skilling current employees through training programs and redesigning work processes providing a better match between the new job demands and available skills. Since future professions are expected to require the adoption and utilization of cloud technologies, smart devices, cyber security technologies, blockchain technologies, automation systems, and autonomous robotics, employees of tomorrow should be equipped with various skills and technical knowledge for certain activities or processes, and learn how to apply them in a digital economy. This is essential for today's people to survive in the dynamic environment of the near future. A profound change in the natures of jobs and working styles seems inevitable, thus they must adapt to change.

The digital age will also introduce various challenges to cope with which will have a profound effect on engineering as a discipline. The environment in which engineering functions are performed and the tools used to perform these functions will change, leading the engineering profile to go through a metamorphosis. Specialized in various technical and social fields ranging from agriculture to electronics, from construction to public works, engineers transform the theoretical knowledge produced by scientists into practical knowledge that technicians and operators can apply, and they have a significant role in the prosperity of nations. Given the economic constraints, engineers are concerned with introducing new technologies and inventions for the development of future systems or the improvement of existing systems. They also design systems for new application areas combining scientific, technical and economic knowledge. The skill sets of today's engineers will go through a transformation with respect to the requirements of the digital age as the ways and places they work change.

According to the Future of Jobs Report 2018 (WEF, 2018, p.12), the future workforce is expected to have some fundamental skills such as (1) analytical thinking and innovation, (2) active learning, (3) creativity, originality and initiative, (4) technology design and programming, (5) critical thinking and analysis, (6) complex problem-solving, (7) leadership and social influence, (8) emotional intelligence, (9) reasoning, problem-solving and ideation, and (10) systems analysis and evaluation. Some, if not all, of these skills will be crucial for practicing engineering. Correspondingly, the current ABET-EAC student outcomes (as of 2019-2020 academic year) state that students of accredited engineering programs should have an ability to (1) identify, formulate, and solve complex engineering problems by applying principles of engineering, science, and mathematics, (2) apply engineering design to produce solutions that meet specified needs with consideration of public health, safety, and welfare, as well as global, cultural, social, environmental, and economic factors, (3) communicate effectively with a range of audiences, (4) recognize ethical and professional responsibilities in engineering situations and make informed judgments, which must consider the impact of engineering solutions in global, economic, environmental, and societal contexts, (5) function effectively on a team whose members together provide leadership, create a collaborative and inclusive environment, establish goals, plan tasks, and meet objectives, (6) develop and conduct appropriate experimentation, analyze and interpret data, and use engineering judgment to draw conclusions, (7) an ability to acquire and apply new knowledge as needed, using appropriate learning strategies (ABET, 2019). Accordingly, engineers of tomorrow should be educated in a way to gain some essential skills and competencies such as

- A thorough understanding of the human element to be integrated into the system they develop or improve;
- Ability to analyze the facts from a multidisciplinary or an interdisciplinary perspective;
- Ability to manipulate and analyze big data and make inferences using it;
- Ability to communicate and collaborate with people from different disciplines and/or cultures;
- Ability to comprehend and analyze social, economic and environmental effects of engineering decisions (develop systems thinking);
- Ability to design and manage technology, take part in the making of corporate-level research and development and innovation policies;
- Ability to generate novel ideas and solutions that are task-appropriate and effective;
- Ability to work in teams in any environment;
- Command of Information and Communication Technologies (ICT);

- Common knowledge management skills;
- Competence in time and quality management;
- Digital literacy in terms of digital content management and creation, and digital security and privacy;
- Flexibility to learning by collaboration under competition;
- Potential and dynamism for lifelong learning;
- Superior tacit knowledge in their professional field/specialization;
- Vision to establish a learning organization.

As automation and digitalization pervade, the expected role of engineers will shift and the engineering profile will revolutionize in accordance with the requirements on the skills and competencies for new-generation engineering jobs. This will thus require the modernization of education systems through the use of virtual reality, autonomous robots, and similar technologies, adopting innovative teaching and learning methods in engineering education, and also updating course contents with respect to digital transformation.

DESCRIPTION OF THE BOOK

Engineers know that the ultimate goal in the process of designing, constructing and delivering products, processes and/or systems of which they are one of the most fundamental elements, is improving the quality of life and social welfare. They strive to use their productivity and ability to innovate for this purpose. The first engineers of modern times played a great role in the advancement of our civilization. This is a legacy that today's engineers have taken over and will pass onto future generations. As distinct from yesterday, engineers of the future will be more focused on people and society.

In parallel with the developments in science and technology and the increasing significance of knowledge in production systems, the physical and social environment of engineers have changed, which in turn, propelled the transformation of engineering profiles. This transformation is also influenced by the change in the nature of materials, products, systems, processes and people. Engineers in the digital age are thus not only expected to be highly ingenious in the design and development of products but also be equipped with the technological tools, trends and practices with regard to digitalization, autonomization, transparency, availability of real-time information and collaboration and be competent in addressing and solving social issues brought by digitalization. In an environment where devices and technologies are interconnected, engineers of the digital age should be highly flexible to work in remotely connected teams, have an interdisciplinary perspective, and be inclined to self-learning and self-improvement. When needed, they should be involved in research and development activities and be competent in technology and innovation management.

Education is a powerful means of shaping the future as one of the most influential activities for social and economic development of individuals and societies. It is clear that the emergence of physical and digital technologies requires a revolution that would affect not only the way we practice engineering but also the way we design and deliver engineering education and professional development/lifelong learning of engineers. Therefore, designing and developing curricula in line with the requirements of future engineering jobs, competencies and skills will be essential for sustainable success. In addition, the use of emergent technologies offers unique and sophisticated solutions that will allow us to achieve improved results in teaching and learning. Subsequently, teachers/educators of engineering education will need to gain relevant competencies such as teaching in virtually enhanced learning environments, incorporating instructional technology to support the learning process, and developing innovative teaching strategies.

The digital transformation will bring about the rethinking and redesigning of engineering education. Thus, there is a need for academic institutions offering engineering degree programs to shift from a traditional way to a more digitized way of designing and delivering the learning and teaching processes and to discuss and act upon how to innovate the engineering education within global, societal, economic and environmental contexts.

The transformation in education requires developing complex, and adaptive learning systems that evolve and connect students' learning needs with the current and novel learning practices, tools and technologies. Designing and delivering such learning systems for the digital era would improve efficiency and effectiveness of engineering education and reduce achievement gaps in student outcomes. It would also help bridging the gap between digital competencies required in the industry and engineering skills developed in academia.

A glance at the relevant literature reveals that different instructional approaches (e.g. project-based learning, problem-based learning, experiential learning, team-based learning, blended learning, game-based learning, self-regulated learning) and educational technologies (e.g. distance education, mobile learning, remote laboratories, virtual laboratories, online learning, e-learning, MOOCs) have been actively used to support engineering education. There is an ongoing debate on how to design the most efficient learning system that would align students' interests and their academic needs with diverse learning resources and networks available. It is obvious that the digital transformation affects the whole curriculum, such that the addition of a single course or a module would not be adequate. However, with the use of advanced educational technologies and instructional approaches, it is possible to integrate the most fundamental aspects of digitalization into the engineering curricula.

For instance, cyber security comes into the picture as a challenge in the digital age, and therefore the curricula of computer science/engineering programs should be revised to provide their students with the ability to address and overcome possible issues brought by this challenge. Similarly, as learning environments become more vulnerable to technology manipulation, higher education institutions will need to better support academic integrity focusing on digital ethics. Higher education institutions are not advised to teach their students ethics solely via coursework. This issue will consequently gain so much importance in the digital age that higher education institutions will need to develop a novel approach for their students to gain a sufficient understanding of digital ethics and cyber behavior. Another example would be the prevalence of new teaching and learning environments such as smart factory labs in which theory and practice are brought together. Industrial, mechanical and electrical engineers can particularly benefit from these innovative environments where they can, before carrying out their projects, design systems using simulation and improve them thereafter with the help of virtual and/or augmented reality. Collaboration with the industry to introduce real-world problems to students will be a great necessity at this point.

ORGANIZATION OF THE BOOK

The book is organized around the issues that have been highlighted and the answers to the following questions: What are the new advancements that would lead to the digital transformation in engineering education? What should be done to ignite this transformation in the first place? What are the new/current paradigms in engineering education and how can we enrich it?

The flow of the chapters in this book is logical and revolves around the design, delivery and/or experience building of engineering education for the digital era. The book is thus divided into two sections, (1) design and (2) delivery and experience, spreading over 11 chapters. The chapters in the first section (Chapters 1-5) focus on defining the educational, organizational, and technological requirements and the design of educational systems, whereas the second sections (Chapters 6-11) focuses on delivering the expectations from engineering education by embedding tools and strategies such as collaborative technologies, challenge-based learning, remote laboratories, and serious games in the engineering curriculum.

Chapter 1 presents a systematic approach to developing engineering education curriculum that adapts to global developments by practicing the philosophy of continuous improvement.

Chapter 2 explores the current state of emerging technologies and educational requirements in engineering education in relation to the developments brought by Industry 4.0.

Chapter 3 introduces two interesting topics: The first one is the development of a course using the ADDIE (Analysis, Design, Development, Implementation, and Evaluation) instructional systems design framework, whereas the second one is the introduction of the TERR (Train, Evaluate, Reflect and Revise) model which acts as a guideline in professional development training.

Chapter 4 presents a literature mapping of studies on educational data mining which is concerned with developing methods for exploring the unique and large-scale data that come from educational settings.

Chapter 5 suggests the use of Management by Values (MBV) in educational organizations to build a more qualified educational environment, to educate highly qualified graduates, and to have highly satisfied faculty by acknowledging shared values.

Chapter 6 discusses the use of collaborative technologies in engineering education and how to design such environments.

Chapter 7 illustrates how to effectively implement the Challenge-Based Learning (CBL) approach in a Higher Education Institution in the UK.

Chapter 8 discusses the experience of the Federal University of Santa Catarina in Brazil with the VISIR+ project, which is an international collaboration project for the dissemination of the remote laboratory VISIR tool to support teaching the theory and practice of electrical and electronic circuits.

Chapter 9 examines a supply chain serious game called “The Fresh Connection” and discusses how it can improve the learning process of engineering students at the university and engineering professionals in the business world.

Chapter 10 merges the use of gamification and serious games for an engineering course, presents the curriculum design for such a course via an application to Game Theory course offered in the Industrial Engineering program at Istanbul Technical University.

Chapter 11 highlights the influence of smartphones and tablets in our society. The design of mobile applications to facilitate student life in the university should fulfill the expectations of the users with respect to several dimensions such as refreshment, life, education, facility, integration with other applications, and transportation to better facilitate their students’ campus life.

CONCLUSION

Not long after the term "Industry 4.0" was coined, a discussion on what kind of new jobs this revolution could create and which existing jobs it would put on the shelf has begun. What kind of skills and competencies would this new era require? More importantly, since it would not be possible for us to provide these skills and competencies with the current practices of education and training, what kind of path would we need to follow in the higher education of future employees so that their skill sets would match the requirements of these jobs?

The influence of this transformation on the engineering profile of tomorrow and, of course, on engineering education that will enhance it is particularly interesting since engineers are expected to find employment in a wide range of areas as a result of various challenges brought about by the digital era that will require the design and construction of new systems or processes or the improvement of existing ones. Thus, engineers of tomorrow should be equipped with numerous skills and competencies that can help them cope with these challenges. This is only possible with the improvement of currently offered engineering education and the development of novel learning and teaching methods.

This book is intended for academics and researchers, educational policymakers, educational planning professionals, educational software developers, undergraduate and graduate students all around the world that are interested in the current knowledge and practice in engineering education of the digital era.

Şeyda Serdar Asan
Instanbul Technical University, Turkery

Erkan Işikli
Instanbul Technical University, Turkey

REFERENCES

ABET. (2019). *ABET Criteria for Accrediting Engineering Programs, 2019 – 2020, Criteria 3 - Student Outcomes*. Retrieved February 20, 2020 from www.abet.org/accreditation/accreditation-criteria/criteria-for-accrediting-engineering-programs-2019-2020

Kagermann, H., Wahlster, W., & Helbig, J. (2013). *Recommendations for Implementing the Strategic Initiative, INDUSTRIE 4.0*. Frankfurt: National Academy of Science and Engineering.

Lu, Y. (2017). Industry 4.0: A survey on technologies, applications and open research issues. *Journal of Industrial Information Integration*, *6*, 1–10. doi:10.1016/j.jii.2017.04.005

Pfohl, H.-C., Yahsi, B., & Kurnaz, T. (2015). The Impact of Industry 4.0 on the Supply Chain. *Proceedings of the Hamburg International Conference of Logistics (HICL)*, 32–58.

Schwab, K. (2017). *The Fourth Industrial Revolution*. New York: Crown Business.

van Laar, E., van Deursen, A., van Dijk, J., & de Haan, J. (2019). Determinants of 21st-century digital skills: A large-scale survey among working professionals. *Computers in Human Behavior*, *100*, 93–104. doi:10.1016/j.chb.2019.06.017

Veile, J. W., Kiel, D., Müller, J. M., & Voigt, K. I. (2018). How to Implement Industry 4.0? An Empirical Analysis of Lessons Learned from Best Practices. In Organizational, and Environmental Determinants, International Association for Management of Technology (IAMOT) Conference, Birmingham, UK.

WEF. (2016). The Future of Jobs Report 2016. *World Economic Forum*. Retrieved January 6, 2020 from http://www3.weforum.org/docs/WEF_Future_of_Jobs.pdf

WEF. (2018). The Future of Jobs Report 2018. *World Economic Forum*. Retrieved January 6, 2020 from http://www3.weforum.org/docs/WEF_Future_of_Jobs_2018.pdf

Xu, L. D., Xu, E. L., & Li, L. (2018). Industry 4.0: State of the art and future trends. *International Journal of Production Research*, *56*(8), 2941–2962. doi:10.1080/00207543.2018.1444806

Acknowledgment

We would like to acknowledge the help of all the people involved in this project. Without their support, this book would not have become a reality.

First of all, we would like to thank Pao-Nan Chou from The National University of Tainan, Katarzyna Grzybowska from Poznan University of Technology, Alexander Pohl from Qualityminds and Seçkin Polat from Istanbul Technical University for their time, expertise, and valuable comments and insights. We appreciate their efforts in seeing this project to completion.

Our sincere gratitude goes to each author who contributed their time and expertise to this book. We thank them for their timely submissions as well as for their willingness to revise and resubmit their work based on reviewer and editor feedback.

We also wish to acknowledge the valuable contributions of the reviewers regarding the improvement of quality, coherence, and content presentation of chapters. Most of the authors also served as referees; we highly appreciate their double task.

Şeyda SerdarAsan
Istanbul Technical University, Turkey

Erkan Işıklı
Istanbul Technical University, Turkey

Section 1

Chapter 1
Outcome-Based Curriculum Design for New-Generation Engineers:
A Case Study From the ITU Metallurgical and Materials Engineering Department

Ayşe Kiliç
Istanbul Technical University, Turkey

Ìsmail Yilmaz Taptik
Istanbul Technical University, Turkey

ABSTRACT

In this study, the stages of change in the curriculum of ITU Metallurgical and Materials Engineering Department (MME) during the ABET accreditation process are summarized and evaluated in terms of both course contents and assessment-evaluation procedures. Improvements in the curriculum design have been carried out within the framework of continuous development, which is one of the essential criteria of today, and then assessment and evaluation methods have been developed, expecting the students become more equipped in their professional life. First of all, a discussion on the characteristics of the Information Age and ABET EAC Student Outcomes (SOs) is provided, followed by the presentation of ITU MME curriculum, conveying the changes in a chronological order. Subsequently, the continuous improvement cycle of ITU MME and the measurement and assessment stages for each outcome are explained. In conclusion, the new ABET EAC SOs are given with a comparison and a final evaluation.

DOI: 10.4018/978-1-7998-2562-3.ch001

INTRODUCTION

In addition to the development and innovation movements of the 21st century, the concept that stands out especially in science & technology and also in almost all areas, is the Information Age that enables learning environment and the development of new educational technologies for everyone. Information age (informatics century), which is considered to be the most important event that humanity has seen after the Industrial Revolution, is clarified as the new period of social and economic organization during which (a) knowledge is the principle source for production, (b) information generation and transmission are widespread, (c) information generation and distribution is largely carried out by employees and (d) change and development are unavoidable through continuous learning and information. Considering the dynamic and sustained formations of globalization and information society, the change in education has become compulsory as it is in the economic, social, political and cultural fields. In this context, a number of different policies and practices have been brought out in the developed and developing countries, such as systematic reforms in education systems, modern institutions, provision of modern teaching tools and equipment, increasing professional knowledge and levels of instructors, improvement of teaching-learning process with changes in managerial sense. Together with the information society, there have been changes in the subjects such as; guiding and instructor trainer, learning student actively at team working environment, instructional-management leadership-based leadership, individual research-based learning method, variable education programs, organizational learning and multi-dimensional conceptual learning criterion. The universities, which have the task of producing knowledge and serving humanity in different disciplines, have become obliged to bring a new perspective to education as required by our age and to use new methods and criteria in order to fulfill their functional duties. When this situation is considered in terms of engineering education, it can be seen clearly since 2000s that graduates of the future engineering programs have become compelled not only to know the fundamentals of science, mathematics, and engineering fields but also to have social skills that will be effective in their professional life. These features also include communication, multi-disciplinary teamwork, problem solving, self-assessment, change management and lifelong learning skills, as well as awareness of social and ethical phenomena related to the engineering profession.

Similarly, ABET has adopted Engineering Criteria2000 in 1997 and since then this has been serving as one of the benchmarks of quality assurance of engineering education world-wide. Among those criteria, achievement of the student outcomes (SO) has been the primary focus of the quality assurance and accreditation requirements for engineering education. According to ABET, Program Educational Objectives (PEO) are "broad statements that describe what graduates are expected to attain within a few years of graduation". Putting "what is learned by the students" to the center of engineering education, ABET EAC defines the SOs as "what students are expected to know and be able to do by the time of graduation. These relate to the knowledge, skills, and behaviors that students acquire as they progress through the program".

Considering the worldwide related issues mentioned above, ITU, has been the focus point in this study in terms of both incorporating leading engineering higher education institutions from various disciplines in its body and also serving as an example as being able to putting itself in a rapid change under fluctuating conditions. One of the primary objectives of the study is to present the systematic approach of adaptation process against the changes from the beginning to the present, and by this to reflect the logic of thinking and progress. In this way, in addition to experience sharing, it will be also shown that the progresses have been conducted in parallel with the global developments approach. As having the awareness that universities have an inevitable duty in the reconfiguration period of education, one of the other objectives is putting an emphasis on the necessity of new graduates have to be equipped with different infrastructure and features at the starting point of their professional life and also revealing this fact by using various assessment tools. It should be also emphasized that these tools are also subjected to continuous improvement process upon requirements.

To summarize, in line with all the objectives, taking the ITU Metallurgical and Materials Engineering (MME) department as a case study, in this study, the stages of change in the curriculum and education perspective of ITU MME Department during ABET Accreditation process are summarized and evaluated in terms of both course contents and assessment-evaluation procedures. Improvements in the curriculum design have been carried out within the framework of continuous development which is one of the essential criteria of today, and then assessment and evaluation methods have been developed and the students are expected to become more equipped in their professional life.

This chapter consists of 4 sections. In the background section, Information age characteristics and ABET EAC Student Outcomes (SOs) are discussed. In Section 3, the curriculum of ITU MME is introduced and the changes are conveyed in a chronological order. After that, the continuous improvement cycle of ITU MME and the measurement & assessment stages for each outcome are explained. In Section 4, the new ABET EAC SOs are given and afterwards comparison and final evaluation are executed.

BACKGROUND

Information Age and Engineering Education (or 21st Century Engineers)

The relationship between information and communication is one of the most crucial and critical phenomena in human life evolution, especially in terms of learning period. We can see different ways of their connection through the elapsed time in such a chronological order; firstly spoken words, initial letters in history, printed books pioneering a new era, the beginning of the information age that revolutionizing communication area, global effect of web technology on communication, etc (Lenschow et al, 1998).

Based on the fact that networks are the basic pattern of life, it is possible to say that communication and social life networks have become the most efficient organizational form due to their prominent features such as flexibility, scalability and survivability against the current technological changes. The technological development that further unleashed the power of network was the transformation of information and communication technologies that emerged in the 1940-1950s with the evolution of the microelectronics industry. This was the foundation of a new technological paradigm which became stronger in the 1970s, primarily in the United States, and rapidly spread out throughout the world. This new era is called as Information Age and descriptively characterized by Manuel Castells (2004).

The Information Age that has the most profound impact on the relationship between information and communication in the near future and also called as the most significant event that mankind has ever seen since Industrial Revolution. Information age focuses on computer technology and data conversion, storage, retrieval and transmission. With the information age, the pace of explosion of knowledge and the ease of access to data are impressive. From now on, more people can achieve data much quickly and in different ways. Also there is an unprecedented increase in the quantity of data and expertise (Rosebrough et al, 2011).

The OECD (Organization for Economic Co-operation and Development) has recently been conducting research and studies in the field of education and learning related to the qualifications that should be gained by the students (Griffin et al, 2012). Supportive results were found in the study conducted by Autor et al (2003). It shows that important shifts have occurred in the workforce structure since 1960 and in this context, “routine” and “manual” tasks are being tremendously replaced speed by “abstract” tasks that require different types of soft and hard skills (**Figure 1).**

Figure 1. Trends in job tasks (Autor et. al., 2003)

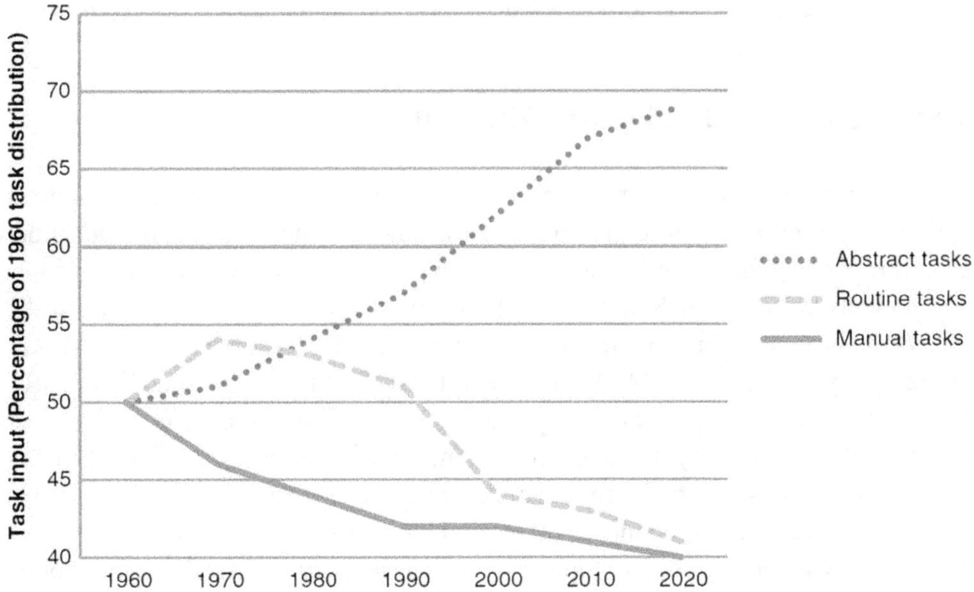

As stated in the report published by the World Economic Forum (2018)., there is a rapid change and differentiation in terms of job descriptions and categories. In this regard, universities have a lot of responsibility in terms of the adaptation of education part to this process (Annand et al, 2007; Kastenberg et al, 2006; Shauman et al, 2002; Doherty et al, 2005; Belski eta l, 2016).

The expectations of the students from the learning environment are also variating. For instance, contrary to the topics given by the professor with typical teacher-based logic, the students want to research and work on topics that overlap with their own interests. Besides, they also have requests for resource support in their areas of interest (Peat et al, 2005).

Similar demands come from industry. These changes and requests are in parallel with technological changes, especially those in the field of information and technology.

The main reason of the change is considered to be information technology and there are many studies related to the effects, advantage and disadvantages of information technology (Catalano et al, 1999; Sutherland et al, 2004).. Change occurs regardless of what name it is called, and more importantly, there is a fresh degree of change. We have to adapt to the changes but much more essential issue is the way that we configure ourselves through the concept of lifelong learning. It means it should be comprised through active life.

In a similar logic to all of the mentioned above, ABET (Accreditation Board for Engineering and Technology) adopted set of standards called as "Engineering

Criteria 2000 (EC2000). In the content of EC2000, the skills and features that graduate student should acquire are listed. Detailed explanations are given in the next and following sections.

The ABET Accreditation and EC2000

ABET was founded in 1932 as the "Engineers' Council for Professional Development (ECPD)" that serves as "an engineering professional body dedicated to the education, accreditation, regulation and professional development of engineering professionals and students in the United States" (ABET, 2019). The first evaluation actions by the ECPD were executed in 1936, resulting in accreditation of a number of engineering degree programs in the USA. By the year 1980, ECPD was renamed as ABET highlighting the importance of quality assurance and accreditation from their aspect (AET, 2019).. The agency started evaluating international programs outside the USA for substantial equivalency by the early 90's and currently it accredits programs in the USA and in 30 more countries all around the world, focusing on four domains of "applied & natural science, computing, engineering, and engineering technology" (ABET, 2019).

ABET EAC's EC2000 (Engineering Criteria2000) has been adopted in 1997, and since then has been serving as one of the benchmarks of quality assurance of engineering education world-wide. Among those criteria, achievement of the student outcomes (SO) has been the primary focal point of the quality assurance and accreditation requirements for engineering education sought by the ABET EAC. Putting "what is learned by the students" to the center of engineering education, ABET EAC defines the SOs as "what students are expected to know and be able to do by the time of graduation. These relate to the knowledge, skills, and behaviors that students acquire as they progress through the program". ABET EAC's criteria for accreditation of engineering programs are identified under 2 sections (ITU, 2019).

- General Criteria: All programs that apply for accreditation from the Engineering Accreditation Commission of ABET are obliged to show that they satisfy all of the following General Criteria for Baccalaureate Level Programs
- Program Criteria: Each program have to satisfy Program Criteria that is specific to the discipline.

General Criteria for Baccalaureate Level Programs are given as a list (ABET, 2019).

- Criterion 1. Students
- Criterion 2. Program Educational Objectives
- Criterion 3. Student Outcomes
- Criterion 4. Continuous Improvement
- Criterion 5. Curriculum
- Criterion 6. Faculty
- Criterion 7. Facilities
- Criterion 8. Institutional Support

ABET Accreditation in Istanbul Technical University

Istanbul Technical University (ITU), which has a history of almost 250 years, is Turkey's first engineering school and also bears the distinction of being one of the leading engineering higher education institutions in the county (ITU, 2019).. Currently, as being located at the central points of Istanbul with 5 city-campuses and housing totally 14 faculties and 86 undergraduate programs, ITU serves to nearly 25,000 undergraduate students out of 40,000 total (undergraduate & graduate) students. The main targets of ITU are (ABET, 2019).

- Education targeting change and improvement
- Outcome-oriented, interdisciplinary research to be beneficial for society
- Effective cooperation in international relations
- Versatile, effective and sustainable university – industry cooperation (UIC)
- Strengthening the ITU perception in public; a participatory and transparent governance and with increasing own revenue

Considering those numbers and facts, providing a quality education and assuring its sustainability becomes indispensable. Accordingly, through the end of 1990's, internationally recognized quality assurance and accreditation–related activities had a jump-start at ITU. Since then, the institute has been leading the field in the country in terms of number of internationally accredited programs: currently, 25 of the engineering undergraduate programs at ITU are accredited by the ABET EAC (Engineering Accreditation Commission) (ABET, 2019).

ABET establishes a six-year cycle of scheduled review for each program. ITU started up its first application for ABET accreditation in 2004 and during the preparatory phase, the restructuring and accreditation activities throughout the whole university are carried out under the framework of EC2000.

ITU Metallurgical and Materials Engineering (MME) Undergraduate Program is one of the leading departments in terms of quality and accreditation studies and has been accredited by ABET since 2004. Due to the changing circumstances and accreditation system requirements, the undergraduate education curriculum of MME has experienced many modifications within the scope of continuous improvement and sustainability. In the following section, ABET accreditation period will be described specific to ITU MME Undergraduate Program.

The ABET History of ITU Metallurgical and Materials Engineering Program and Restructuring Period of the Undergraduate Curriculum

The Education Language of the Program

Reconfiguration of the undergraduate education with the ABET approach has started in 1996. In the beginning, ITU University Senate has taken 2 important decisions:

- Teaching 30% of the courses in all of the programs' curriculum in English
- Reducing the course loads to 150 credits in the 8-semester undergraduate curriculum

Firstly one of the important decision taken is that our students should know another language, preferably English in order to be able to compete in the international arena as well as to follow global developments easily. Thus before the ABET audit 2003, the university and the programs adopted to teach 30% of the courses in English. After ABET audit 2009, ITU University Senate, has given another decision about the education language. As a result of discussions and exchange of ideas with the students, alumni, employers and active stakeholders, education of ITU programs became either 100% English or 30% English since the 2010-2011 academic calendar. 100% English education of ITU MME Undergraduate program has been started with parallel to the 30% English education option since 2011-2012 academic year.

Newly enrolled students must take the Proficiency Exam prepared by the ITU School of Foreign Languages to prevent problems while attending the English classes. Students' test score should be 60 or above out of 100 (assumed to be equivalent to 513 on TOEFL PBT+TWE, 65 ON TOEFL IBT, 55 on PBT or 6 on the IELTS taken within the past two years) to be able to start education in their own program (ITU, 2019). The ones having score below 60, should take the Placement Exam to be placed in the appropriate level in the prep program. This mandatory one-year intensive language (English) teaching program (Prep class) has been implemented since 1997-1998 academic year. There are 4 levels in the programs and each level (A-level: Upper-Intermediate, B-level: Intermediate, C-level: Pre-Intermediate, D-level: Beginner) the total number of educations hours differ depending on the language proficiency. Apart from these, ITU offers elective, free of charge language courses in German, French and Japanese for its students. The organization of these courses are conducted at the beginning of each term (at different 3 levels: beginner, intermediate and advanced). Students are given a certificate after completing these courses.

Restructuring Period of the Undergraduate Curriculum

Another most essential decision taken to be implemented at the whole university is that changing course syllabuses within the framework of ABET criteria. ITU MME Undergraduate program provides its students to choose their option after 3rd grade. In 2004, MME Program, considering the industrial needs, redesigned its undergraduate education offering students 3 options: Metallurgy, Materials and Ceramic. As there were none enrolled students to the ceramic option between the years 2004-2010, after consulting with the stakeholders, the academic council of the program has decided to close up Ceramic option and continued its education under 2 options as; metallurgy and materials. At the beginning of 2011-2012 Academic year, two-option education; metallurgy and materials, has been modified to one-option called as Metallurgical and Materials Engineering with reorganizing the course syllabi.

This mentioned reorganization of the curriculum has been executed as a result of the surveys carried out on students, employers and also making criticism with the staff. Since 2011-2012 Academic year, MME Undergraduate Curriculum consists of 150 course credits (240.5 ECTS Credits). Courses in the curriculum can be classified under 4 main groups (there are subgroups as required and elective courses):

- Basic Science Courses (BS)
- Engineering Science Courses (ES)
- Engineering Design Courses (ED)
- General Education Courses (GE)

In first 2 years (1st to 4th Semesters), basic science (42 hours, 38 credits) and engineering science courses (48 hours, 42 credits) should be taken by the students as they are required in providing a solid basis for the following engineering classes. Engineering design based mandatory and elective courses are spread over 5th to 8th semesters with the changing number and diversity (43 hours, 36.5 credits). There are more than 100 options for general education courses and students have to complete 29 credit hours with them throughout their whole undergraduate education.

Through all ABET Accreditation cycles, there is no difficulty in acquiring hard skills to the students. Starting from the fact that one of the most essential requirement is the acquisition of soft skills, a different structuring process was introduced in the curriculum. There are 3 project-based lectures (MET 353E: Design Principles and Material Selection, MET 348E: Quality Engineering, MET 435E: Problem Solving techniques, Project Management and Innovation) at the successive semesters (5th, 6th and 7th Semesters) with the aim of both enabling students to express themselves orally and verbally and also developing the teamwork skills. These courses are also a preparation phase for Graduation Design project. Improvements have been conducted to the Graduation Design Project course that students are taking in the 8th semester (spring term). According to the University Senate Decision, total course credits are increased with offering the Graduation Design Project in two consecutive sections as "Engineering Design-I" and "Engineering Design-II" in fall and spring semesters, respectively. At the end of these courses, students' ability of making presentations, writing reports, project management and working in teams show a great increase.

Another important innovation is the courses that are taken by the students at the 1st and 6th Semester as "Academic Advising" and "Career Advising". Academic advising is a training/education process that helps to bring meaning and purpose to undergraduate study. Through the undergraduate study, it provides intellectual and personal development of academic achievement and lifelong learning. Academic advising in ITU provides opportunities for students to gain access to education, personal and career goals, realize the abilities needed for academic success, learn how to access the countless resources and services offered by the institution, and establish relationships with one or more academic advisers. Career Advising course is for the students who have taken their professional courses and now at the phase of shaping their career path. They have a great opportunity to have consultancy from academic staff about experience, knowledge and, most importantly, how to proceed according to their interests. Under the favor of the academic consultancy, which has become one of the strengths of ITU, the social relationship part is strengthened and the sense of belonging increases.

ITU MME undergraduate program's education quality derives its strength from the rich infrastructure of its academic staff that all of them have required knowledge, experience and skills for a contemporary teaching in metallurgical and materials engineering field. Also they have both international work experience and also social & scientific relationships with colleagues from national & international institutions. As all of the academic staff members have focused their primary duties on education and teaching, they give a considerable effort in improving the way of their teaching.

All the courses in the curriculum are taught by the programs' academic staff. There has been a large team helping the academic staff and consisting of young researchers, research assistants, volunteer seniors, graduate students and etc. and programs' academic infrastructure acted with lifelong learning logic and expanded its staff gradually.

Not only by changing course syllabuses and curriculums of the undergraduate education, but also within the framework of ABET EC2000, ITU MME has redefined its mission, vision, program educational objectives and outcomes during the years

2000-2002 and started to implement effectively and adopted the philosophy of continuous improvement within all the cycles. In the next section, program educational objectives and student outcomes of MME Program will be explained in detail with their assessment tools used during the continuous improvement cycle.

Continuous Improvement Cycle of ITU MME Undergraduate Program

Within the ABET accreditation structure, outlined as general criteria in Section 2.b, each program should set it program educational objectives and student outcomes. As the last audit of ABET was in 2015, the continuous improvement cycle steps that will be explained in this section, have been conducted in accordance to the criteria valid at that time.

Both of the PEOs and SOs are shared with public through the program's website (ITU, 2019). and also all of the students are informed about them via written document at the introductory course namely; MET 101E-Introduction to Metallurgical & Materials Engineering and Engineering Ethics.

ITU MME Undergraduate Program Educational Objectives

According to ABET (2015-2016 Criteria for Accrediting Engineering Programs), program educational objectives are defined as "broad statements that describe what graduates are expected to attain within a few years of graduation. Program educational objectives are based on the needs of the program's constituencies" (ABET, 2019).

Program Educational Objectives (PEO) of ITU MME Undergraduate Program that provides its graduate students with, are listed below:

PEO 1: A sound basis and application skills in mathematics, physics, chemistry, physical chemistry and basic engineering,

PEO 2: Knowledge in the use of techniques and equipment required for modern engineering applications and the ability to utilize this knowledge in design, application and communication,

PEO 3: The skills required to characterize structure-property-processing and performance of materials, and the metallurgical production parameters with standard or self-designed experimental techniques and to interpret the results,

PEO 4: The knowledge of the basic concepts of metallurgy and materials science and engineering and the ability of materials evaluation within the framework of structure-property-process-performance relations,

PEO 5: Intensive knowledge in the production of metallic and non-metallic materials from primary and secondary resources and in the processes and technologies related to processing, and protection of these materials, and the ability to apply this knowledge in the application and development,

PEO 6: The tools necessary to define engineering problems, choose and design suitable material, system, product, and process, and to transform these into projects which are economically sound whilst taking into consideration the conservation of the environment and quality of the product,

PEO 7: Professional and ethical responsibilities in following and evaluating contemporary and social developments, oral and written communication skills, a teamwork environment, and the desire to continuously learn and progress.

While setting up Program's educational objectives, there were no national program criteria to build the Program's program educational objectives and student outcomes. The international criteria in the field of engineering set by Minerals, Metals and Materials Society (TMS) in the USA has been a guide (TMS, 2019). These criteria are believed to be in harmony with the program educational objectives and student outcomes.

The Minerals, Metals and Materials Society's program criteria are given below:

- **Criteria 1:** Ability to apply advanced science and engineering to materials systems implied by program modifier (ceramics, metals, polymers, composite materials, etc.)
- **Criteria 2:** Integrated understanding of structure, properties, processing and performance of materials systems
- **Criteria 3:** Ability to solve materials selection and design problems
- **Criteria 4:** Ability to utilize experimental, statistical and computational methods consistent with program educational objectives.

Important concepts listed by TMS are: Structure, Properties, Experimental Design/Analysis, Process, Cost/Performance, Quality/Environment, Product and Process Design.

Table 1 illustrates the relationship between MME program educational objectives and TMS Program criteria. Table 2 shows the relationship between MME program educational objectives and key concepts defined both by the program and the TMS criteria.

Table 1. The relationship between the Metallurgical and Materials Engineering (MME) Undergraduate program's educational objectives and TMS program criteria

MME Program Education Objectives		TMS Program Criteria		
	1	2	3	4
I		x	x	x
II	x	x	x	x
III	x	x	x	x
IV	x	x	x	
V	x	x	x	
VI	x	x	x	x
VII			x	

Table 2. The relationship between the Metallurgical and Materials Engineering (MME) Undergraduate program's educational objectives and key concepts defined both by the program and the TMS

	MME Key Concepts						
MME Program Education Objectives	Structure	Properties	Design Experiment Analyze Data	Processing	Cost/ Performance	Quality/ Environment	Design Process or Product
I			x				x
II	x	x	x	x			x
III	x	x	x				x
IV	x	x		x			x
V	x	x		x			x
VI	x	x	x	x	x	x	x
VII						x	x

Curriculum courses align with the program educational objectives. Table 3 shows the relationship between the program educational objectives and programs' courses. Orange color (Course codes between KIM 101E-BIL108E) represents the "Basic Science Courses". Green colored courses (Course codes between ELK221E-MET 326E) are "Engineering Science Courses" and blue color (Course codes between MET 337E-MET476E) represents the "Engineering Design Courses".

Table 3. The relationship between the Metallurgical and Materials Engineering (MME) Undergraduate program's educational objectives and programs' courses.

Course Code & Course Name		PEO I	PEO II	PEO III	PEO IV	PEO V	PEO VI	PEO VII
KIM 101E	General Chemistry I	x						
KIM 101EL	General Chemistry I Lab	x		x				
MAT 103E	Mathematics I	x						
FIZ 101E	Physics I	x						
FIZ 101EL	Physics I Lab	x		x				
KIM 203 E	Analytical Chemistry	x						
FIZ 102EL	Physics II Lab	x		x				
FIZ 102E	Physics II	x						
MAT 104	Mathematics II	x						
MAT 201E	Differential Equations	x						
FIZ 201E	Modern Physics	x						
MAT 271E	Probability and Statistics	x						
KIM 205E	Organic Chemistry	x						
BIO 301E	Fundamentals of Biology	x						
BIL 101E	Introduction to Computers and Information Systems		x	x				x
RES 107E	Technical Drawing		x					
BIL 108E	Introduction to Scientific & Engineering Computing		x	x				x
ELK 221E	Fundamentals of Electrical Engineering	x		x				
MET 101E	Introductions to Metallurgy & Materials Engineering & Engineering Ethics	x						x
MET 215E	Fundamentals of Metallurgical Thermodynamics	x			x			
MET 213E	Fundamentals of Materials Science			x	x			
MET 214E	Fundamentals of Solution Thermodynamics	x			x			
MET 224E	Phase Equilibrium Diagrams				x			
MET 248E	Mass and Energy Balance	x						
MET 244E	Static and Dynamic Strength of Materials	x	x					
MET 246E	Materials Physics	x			x			
MET 228E	Materials Chemistry	x			x			
MET 248E	Mass and Energy Balance		x		x			
MET 313E	Chemical Metallurgy I			x	x	x		
MET 315E	Fundamentals of Physical Metallurgy			x	x	x		
MET 317E	Fundamentals of Transport Phenomena		x					
MET 326E	Chemical Metallurgy II			x	x	x		
MET 337E	Materials Characterization Meth.		x	x	x			
MET 353E	Design Principles & Materials Selection for Engineering Applications		x	x			x	x

continued on following page

Table 3. Continued

Course Code & Course Name		PEO I	PEO II	PEO III	PEO IV	PEO V	PEO VI	PEO VII
MET 344E	Mechanical Properties of Materials			x				
MET 346E	Modelling and Simulation of Metallurgical and Materials Processing		x	x			x	
MET 348E	Quality Engineering		x				x	x
MET 364E	Extractive Metallurgy Laboratory		x	x				
MET 435E	Problem Solving Techniques, Design and Project Management		x				x	x
MET 437E	Principles of Corrosion and Corrosion Protection				x	x		
MET 439E	Engineering Polymers				x	x		
MET 442E	Fundamentals of Composite Materials				x	x		
MET 453E	Fundamentals of Ceramic Materials				x	x		
MET 455E	Materials Processing Laboratory		x	x				
MET 492E	Design Project		x				x	x
MET 444E	Iron and Steel Materials in Engineering Applications				x	x		
MET 446E	Environment & Ethics in Metallurgical & Materials Processing						x	x
MET 366E	Energy Materials				x			
MET 368E	Production Techniques of Metallic Powder				x	x		
MET 374E	Process Metallurgy				x	x		
MET 376E	Heat Treatment of Metals				x	x		
MET 457E	Semiconductor Materials				x			
MET 475E	Powder Metallurgy				x	x		
MET 477E	Intro. To Electrometallurgy				x	x		
MET 479E	Iron and Steel Making				x	x		
MET 483E	Introduction to Non-Destructive Testing of Materials			x	x			
MET 485E	Ceramic Manufacturing				x	x		
MET 487E	Deformation Processes of Materials				x			
MET 462E	Metallic Nanoparticles: Production & Characterization				x	x		
MET 464E	Principles of Surface Treatment				x	x		
MET 468E	Welding Technology and Metallurgy				x	x		
MET 472E	Casting Processes				x	x		
MET 474E	Extractive Metallurgy of Non-Ferrous Metals				x	x		
MET 476E	Introduction to Technical Ceramics				x	x		

ITU MME Undergraduate Program's Student Outcomes

Student outcomes (SOs) assist the MME Program to achieve its educational objectives. The definition for student outcomes by ABET as; "describe what students are expected to know and be able to do by the time of graduation. These relate to the

Figure 2. The Continuous improvement loop of the ITU MME Undergraduate program

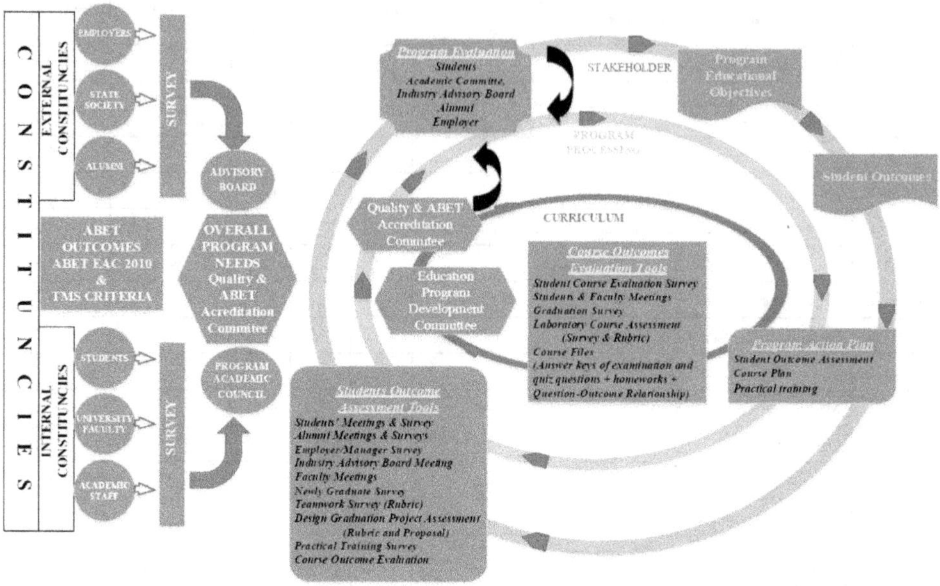

Table 4. Assessment Schedule and responsibilities of the continuous improvement loop

PROCESSING	SCHEDULE	RESPONSIBILITY
Curriculum Loop	Once a year	Department-Educational Progeam Development Committee
Processing Loop	Every 3 years	Department-Quality and Accreditation Committee, Industry Advisory Board
Stakeholder Loop	Every 5 years	All Program Constituencies

Table 5. Assessment tools for the student outcomes of ITU MME Undergraduate program

Assessment Tool	Student Outcome	Schedule	Aim	Constituency	Responsible
The course evaluation survey	SO (A), SO (B), SO (D), SO (E), SO (F), SO (I)	At the end of every semester	To see course objectives have or have not been met	1st, 2nd, and 3rd Grade Students	Chari, Student, Lecturer, Education Program Development Committee
Course outcome evaluation	SO (A), SO (B), SO (C), SO (D), SO (E), SO (F), SO (G), SO (H), SO (I)	At the end of every semester	To analyze if the course outcomes are achieved	1st, 2nd, 3rd Grade Students	Lecturer, *(Education Program Development Committee, Quality & ABET Acc. Comm*)*
Annual Course assessment meetings with students	SO (A), SO (B), SO (E), SO(G), SO (I)	At the end of each semester	To have different ideas for the development of the course given	1st, 2nd, and 3rd Grade Students	Lecturer; Chari of Education Program Development Committee
Students' meetings and their survey	SO (A), SO (B), SO (C), SO (D), SO (E), SO (F), SO (G), SO (I)	Each year	To follow the students' improvement and determine their problems	1st, 2nd, and 3rd Grade Students	Chari, Students
Industry Advisory Board Meeting	SO (A), SO (B), SO (C), SO (D), SO (E), SO (F), SO (G), SO (H), SO (I)	Twice a year	To see the compliance level of program educational objectives and outcomes for the industries' needs	Industry representatives, employers, alumni	Chari, Education Program Development Committee Advisory Board Members
Faculty Annual Assessment Meeting	SO (A), SO (B), SO (C), SO (D), SO (E), SO (F), SO (G), SO (H), SO (I)	At least twice in a year	To give faculty to evaluate educational objectives, outcomes and department services	Chair Faculty	Chair Faculty
Alumni Surveys fo 1-3-5 years out	SO (A), SO (B), SO (C), SO (D), SO (E), SO (F), SO (G), SO (H), SO (I)	Every 5 years	To see the perceptions of the alumni on MME education after graduation & also to evaluate the level of contribution of their education to their	≤1 years out 1-3 years out 5≤years out	Chair Alumni
Newly Graduate Survey	SO (A), SO (B), SO (C), SO (D), SO (E), SO (F), SO (G), SO (H), SO (I)	Immediately after graduation	By students' point of view to see whether the education they received met the objectives and outcomes determined	Students conducting thei Design Graduation Project Final Presentation	Chair Senior Students Education Program Development Committee, Quality & ABET Acc. Comm
Employer/ Manager Survey	SO (A), SO (C), SO (D), SO (E), SO (F), SO (G), SO (H), SO (I)	Every 3 years	To evaluate the sufficiency of our students in professional life in terms of student outcomes and program educational objectives	Employers and/or managers	Alumni Committee, Industrial Association Committee, Internship committee
Laboratory Course Assessment (Rubric and Survey)	SO (B)	Every Semester	Survey aims to have the suggestions from the students to improve the laboratory courses; Rubric aims to evaluate performances of the students during experiments with specific criteria.	3rd and 4th grade students	Students Lecturer
Practical Training Survey	SO (B), SO (C), SO (D), SO (E), SO (F), SO (G), SO (I)	Every Semester	To understand the level of contribution of practical trainings in education of students	1st, 2nd, and 3rs Grade Students	Chair Students
Design Graduation Project Assessment (Rubric and Proposal)	SO (C), SO (D), SO (E), SO (G), SO (H), SO (I)	At the end of the semester	The Design Graduation Project is aimed to evaluate student's level of knowledge, creativity, and ingenuity as well as his/her ability to present them both orally and in written form.	Design Graduation Project Students	Jury, All Faculty members, Research Assistants, Senior Students Advisory board members (if they attend)
Team Work Rubric	SO (C), SO (D), SO (H)	Every Semester	To understand the sufficiency and success rate of the team work	3rd and 4th Grade Students	Students Lecturers

skills, knowledge, and behaviors that students acquire as they progress through the program" (ABET, 2019).

The SOs of the MME Undergraduate program are listed below (ITU, 2019).

SO (A): Ability to apply the knowledge of mathematics, science, and engineering principles to solve problems in metallurgical and materials engineering (ABET:a)

SO (B): Ability to characterize materials using standard and/or self designed experimental methods and to evaluate the results (ABET:b)

SO (C): Ability to design a system or a process, taking into consideration of the desired specifications, quality, ethics, and environment (ABET:c)

SO (D): Ability to communicate both verbally and in the written form and to take part in, and provide leadership of the teams in the elucidation of engineering problems (ABET:d, g)

SO (E): Ability to define, formulate and solve engineering problems in the development, production, processing, protection, and usage of engineering materials (ABET:e)

SO (F): An understanding of professional and ethical responsibilities (ABET:f)

SO (G): An understanding of current / contemporary issues and impact of engineering solutions in broad cultural, national and global levels (ABET:h, j)

SO (H): A comprehension of the nature of engineering progress closely linked with the development of new materials and production processes. An ability to engage in life-long learning and a recognition of its necessity (ABET:i)

SO (I): Ability to use essential tools and techniques of modern engineering in the development, production, processing, protecting of the existing and new engineering materials. (ABET:k)

Continuous Improvement Loop Diagram

According to the ABET accreditation, "The program must regularly use appropriate, documented processes for assessing and evaluating the extent to which the student outcomes are being attained. The results of these evaluations must be systematically utilized as input for the continuous improvement of the program" (ABET, 2019).. Due this requirement, for the ABET audit 2009 and 2015, the continuous improvement loop diagrams are created and modified (Figure 2).

The continuous improvement loop diagram have been created considering the inputs of the constituency as: Curriculum loop, Processing Loop and Stakeholder Loop. Table 4 gives the details about the assessment schedule and responsibilities.

Table 6. Mapping diagram of the student outcomes after and before the 2019-2020 accreditation cycle.

Student Outcomes after 2019-2020 Accreditation Cycle Student Outcomes before 2019-2020 Accreditation Cycle	**1**	**2**	**3**	**4**	**5**	**6**	**7**
	an ability to identify, formulate, and solve complex engineering problems by applying principles of engineering, science, and mathematics	**an ability to apply engineering design to produce solutions that meet specified needs with consideration of public health, safety, and welfare, as well as global, cultural, social, environmental, and economic factors**	**an ability to communicate effectively with a range of audiences**	**an ability to recognize ethical and professional responsibilities in engineering situations and make informed judgments, which must consider the impact of engineering solutions in global,**	**an ability to function effectively on a team whose members together provide leadership, create a collaborative and inclusive environment, establish goals, plan tasks, and meet objectives**	**an ability to develop and conduct appropriate experimentation, analyze and interpret data, and use engineering judgment to draw conclusions**	**an ability to acquire and apply new knowledge as needed, using appropriate learning strategies**
(a) an ability to apply knowledge of mathematics, science, and engineering	x						
b) an ability to design and conduct experiments, as well as to analyze and interpret data						x	
(c) an ability to design a system, component, or process to meet desired needs within realistic constraints such as economic, environmental, social, political, ethical, health and safety, manufacturability, and sustainability		x					
(d) an ability to function on multidisciplinary teams					x		
e) an ability to identify, formulate, and solve engineering problems	x						
(f) an understanding of professional and ethical responsibility				x			
(g) an ability to communicate effectively			x				
(h) the broad education necessary to understand the impact of engineering solutions in a global, economic, environmental, and societal context				x			
(i) a recognition of the need for, and an ability to engage in life-long learning							x
(j) a knowledge of contemporary issues				x			
(k) an ability to use the techniques, skills, and modern engineering tools necessary for engineering practice.	x	x				x	

Program Educational Objectives are revised in every 5 years by taking into account the results of curriculum and processing loops' assessments.

Table 5 shows the tools used for measuring and assessing the student outcomes. Timing period, aim, constituency and responsible people are also explained. The results of these assessments are analyzed by the initially Quality and Accreditation Committee and later by Academic Council and Advisory Board.

FUTURE RESEARCH DIRECTIONS AND CONCLUSION

Upon the announcement made by ABET, vital changes occurred in the Accreditation Criteria (ITU, 2019). Newly approved criteria is approved to be implemented in the 2019-2020 cycle and will be mandatory for all the programs. "Student Outcomes" section consists the most important changes. Different from the 11 Student Outcomes (a-k), there will be 7 SOs that the programs' graduates should attain. The new SOs are given below:

- An ability to identify, formulate, and solve complex engineering problems by applying principles of engineering, science, and mathematics (1)
- An ability to apply engineering design to produce solutions that meet specified needs with consideration of public health, safety, and welfare, as well as global, cultural, social, environmental, and economic factors (2)
- An ability to communicate effectively with a range of audiences (3)
- An ability to recognize ethical and professional responsibilities in engineering situations and make informed judgments, which must consider the impact of engineering solutions in global, economic, environmental, and societal contexts (4)
- An ability to function effectively on a team whose members together provide leadership, create a collaborative and inclusive environment, establish goals, plan tasks, and meet objectives (5)
- An ability to develop and conduct appropriate experimentation, analyze and interpret data, and use engineering judgment to draw conclusions (6)
- An ability to acquire and apply new knowledge as needed, using appropriate learning strategies (7)

Mapping matrix is created with the newly proposed and the original Student Outcomes to show the differences much more clearly (Table 6).

As can be seen from the details of the study, the variation in the expectations of the developing and changing engineer profile leads to a requirement of a highly planned, systematic and long process in engineering education. At this point, it can be achieved by analyzing the expectations very strategically at the beginning and then reflecting the results to the curriculum. Henceforward, the assessment and evaluation techniques come into prominence. These mentioned techniques were previously conducted through surveys but from now on, survey results has started to be the weakest link in the assessment and evaluation process. In recent years, variable measurement and evaluation methods have been developed especially for the assessment of soft skills. There is no reason of not realizing all of the mentioned shifts as long as the academic staff are open to change.

In this study, some of these examples and the obtained results are presented. For the further study, based on the ABET changes, continuous improvement loop steps will be revised according to the new ABET EAC SOs. This revision stage will comprise the assessment tools, especially the rubrics and survey questions.

REFERENCES

ABET. (2019). *ABET history*. Retrieved from https://www.abet.org/about-abet/history/

ABET. (2019). *ABET accreditation criteria for accrediting engineering programs in 2019-2020*. Retrieved from https://www.abet.org/accreditation/accreditation-criteria/criteria-for-accrediting-engineering-programs-2019-2020/]

ABET. (2019). *ABET accredited program details*. Retrieved from http://main.abet.org/aps/AccreditedProgramsDetails.aspx?OrganizationID=554&ProgramIDs=

ABET. (2019). *Accreditation changes*. Retrieved from https://www.abet.org/accreditation/accreditation-criteria/accreditation-changes/

Annand, D. (2007). Re-organizing universities for the information age. *International Review of Research in Open and Distance Learning*, *8*(3), 1–9. doi:10.19173/irrodl.v8i3.372

Autor, D., Levy, F., & Murnane, R. (2003). The skill content of recent technological change: An empirical exploration. *The Quarterly Journal of Economics*, *118*(4), 1279–1333. doi:10.1162/003355303322552801

Belski, I., Adunka, R., & Mayer, O. (2016). Educating a creative engineer: Learning from engineering professionals. *Procedia CIRP*, *39*, 79–84. doi:10.1016/j.procir.2016.01.169

Castells, M. (2004). Informationalism, networks, and the network society: a theoretical blueprint. In M. Castells (Ed.), *Network Society: A cross-cultural perspective* (pp. 3–45). Cheltenham, UK: Edward Elgar Publishing Limited. doi:10.4337/9781845421663.00010

Catalano, G. D., & Catalano, K. (1999). Transformation: From teacher-centered to student-centered engineering education. *Journal of Engineering Education*, *88*(1), 59–64. doi:10.1002/j.2168-9830.1999.tb00412.x

Doherty, J. J., Hansen, M. A., & Kaya, K. K. (2005). Teaching information skills in the information age: The need for critical thinking. *Library Philosophy and Practice*, *1*(2), 1–9.

Einsenberg, M. B. (2008). Information literacy: Essential skills for the information age. *DESIDOC Journal of Library and Information Technology*, *28*(2), 39–47. doi:10.14429/djlit.28.2.166

Fruchter, R. (2001). Dimensions of teamwork education. *International Journal of Engineering Education*, *17*(4-5), 426–430.

Griffin, P., Care, E., & McGaw, B. (2012). The changing role of education and schools. In P. Griffin, E. Care, & B. McGaw (Eds.), *Assessment and teaching of 21st century skills* (pp. 1–15). New York: Springer. doi:10.1007/978-94-007-2324-5_1

ITU. (2019). *ITU history*. Retrieved from http://global.itu.edu.tr/global-itu/history

ITU. (2019). *ITU vision and mission*. Retrieved from http://global.itu.edu.tr/global-itu/vision-and-mission

ITU Metallurgical and Materials Engineering Undergraduate Program. (2019). *Program educational objectives*. Retrieved from http://mme.itu.edu.tr/en/abet/educational-objectives

ITU Metallurgical and Materials Engineering Undergraduate Program. (2019). *Student Outcomes*. Retrieved from http://mme.itu.edu.tr/en/abet/student-outcomes

ITU School of Foreign Languages. (2019). *Minimum English scores for undergraduate programs*. Retrieved from http://www.ydy.itu.edu.tr/en/to-be-exempt-from-program/

Kastenberg, W. E., Hauser-Kastenberg, G., & Norris, D. (2006). *Proceedings of 36th ASEE/IEEE Frontiers in Education Conference*. IEEE.

Konsky, B. R., Miller, C., & Jones, A. (2016). The skills framework for the information age: Engaging stakeholders in curriculum design. *Journal of Information Systems Education*, *27*(1), 37–49.

Lenschow, R. J. (1998). From teaching to learning: A paradigm shift in engineering education and lifelong learning. *European Journal of Engineering Education*, *23*(2), 155–161. doi:10.1080/03043799808923494

Moncef Nehdi, P. E. (2002). Crisis of civil engineering education in information technology age: Analysis and prospects. *Journal of Professional Issues in Engineering Education and Practice*, *128*(3), 131–137. doi:10.1061/(ASCE)1052-3928(2002)128:3(131)

Peat, M., Taylor, C. E., & Franklin, S. (2005). Re-engineering of undergraduate science curricula to emphasise development of lifelong learning skills. *Innovations in Education and Teaching International*, *42*(2), 135–146. doi:10.1080/14703290500062482

Raju, P. K. (2003). Educating engineers for the information age. In *Proceedings of American Society for Engineering Education Annual Conference & Exposition*, (Session 2558, pp.8459.1-8459.14). Academic Press.

Reigeluth, C. M., Watson, W. R., Watson, S. L., Dutta, P., Chen, Z., & Powell, N. D. P. (2008). Roles for technology in the information-age paradigm of education: Learning management systems. *Educational Technology*, *48*(6), 32–39.

Rosebrough, T. M., & Leverett, R. G. (2011). *Transformational teaching in the information age*. Alexandria, VA: ASCD Publication.

Shuman, L., Atman, C. J., Eschenbah, E. A., Evans, D., Felder, M., Imbrie, P. K., ... Yokomoto, C. F. (2002). *Proceedings of 36th ASEE/IEEE Frontiers in Education Conference*. IEEE.

Sutherland, R., Robertson, S., & John, P. (2004). Interactive education: Teaching and learning in the information age. *Journal of Computer Assisted Learning*, *20*(6), 410–412. doi:10.1111/j.1365-2729.2004.00100.x

TMS. (2019). *University Accreditation ABET*. Retrieved from https://www.tms.org/portal/PROFESSIONAL_DEVELOPMENT/Professional_Development_Resources/University_Accreditation__ABET_/portal/Professional_Development/Professional_Development_Resources/University_Accreditation__ABET_.aspx?hkey=6efc343b-effd-40d3-8cb2-b5c3934981ef

Waddoups, G. L., Wentworth, N., & Earle, R. (2004). Principles of technology integration and curriculum development. *Computers in the Schools*, *21*(1-2), 15–23. doi:10.1300/J025v21n01_02

Wang, Y. (2002). From teacher-centredness to student-centredness: Are preservice teachers making the conceptual shift when teaching in information age classrooms? *Educational Media International, 39*(3-4), 257–265. doi:10.1080/09523980210166710

World Economic Forum. (2018). *The future of jobs*. Retrieved from http://www3.weforum.org/docs/WEF_Future_of_Jobs_2018.pdf

ADDITIONAL READING

Brabazon, T. (2007). *The university of google: education in the (post) information age*. USA: Ashgate Publishing Company.

Castells, M. (2009). *The power of identity: the information age-economy, society, and culture*. Hoboken, New Jersey, USA: John Wiley & Sons. doi:10.1002/9781444318234

Castells, M. (2010). *End of millennium: the information age-economy, society, and culture*. Hoboken, New Jersey, USA: John Wiley & Sons. doi:10.1002/9781444323436

Chandrasekaran, S., Stojcevski, A., Littlefair, G., & Joordens, M. (2013). Accreditation inspired project oriented design based learning curriculum for engineering education. In *Proceedings 2nd International Engineering and Technology Education Conference Ho Chi*, (pp.1-11). Ho Chi Minh City, Vietnam.

Felder, R. M., & Brent, R. (2003). Designing and teaching courses to satisfy the ABET engineering criteria. *Journal of Engineering Education, 92*(1), 7–25. doi:10.1002/j.2168-9830.2003.tb00734.x

Idachaba, F. (2018). Outcome based engineering curriculum design: a system for curriculum streamlining and graduate quality improvement in engineering. In *Proceedings of INTED2018 Conference,* (pp.5888-5893). 10.21125/inted.2018.1396

Liebelt, M., Eglinton-Warner, S., Soong, W., Al-Sarawi, S., Ng, B., Phillips, B., & Sorell, M. (2017). An engineering approach to engineering curriculum design. In *Proceedings of 28th Australasian Association for Engineering Education (AAEE) Annual Conference,* (pp.777-784). Sydney, Australasian Association for Engineering Education.

World Economic Forum. (2016). *New vision for education: fostering social and emotional learning through technology*. Retrieved from http://www3.weforum.org/docs/WEF_New_Vision_for_Education.pdf

KEY TERMS AND DEFINITIONS

ABET Accreditation: Agency that is non-profit and non-governmental that accredits the various programs in higher education such as applied and naturel sciences, computing, engineering, and engineering technology. This provides mentioned programs in college or university meets the quality standards of the profession that the program is preparing its graduates.

Assessment Tools: Techniques or instruments are used with the intention of collecting data for numerous reasons such as evaluating and measuring the outcomes in terms of learning process, educational needs, and skill acquisition of the students.

Continuous Improvement: As improvement is not something that has a starting and an ending, this term indicates the ongoing process of learning, self-improvement, adaptation and growth.

Educational Objectives: Goals that the education process is directed in line with. They provide a common language that makes a linkage between subjects, assessment and grade levels.

Hard and Soft Skills: Hard skills can be defined as the typically quantifiable skills that can be easily defined and evaluated and also the ones you are able to gain through education, courses, training, certificate programs, etc. On the other side, soft skills are interpersonal skills that are more difficult to be defined and assessed such as communication skills, empathy, creative thinking, motivation, teamwork, etc.

Information Age: Historical period that starts in the beginning of 20th century and mainly emphasized by the rapid development of the information technology related issues.

Outcome-Based Assessment: Techniques basically have a deeper concentration on the measurement and evaluation of student`s performance and achievement of success.

Project-Based Learning: A student centered and dynamic educational approach and pedagogy in which students can actively acquire a deeper knowledge in projects by exploring real-world challenges and problems.

Student Outcomes: Knowledge, skills and attitudes that are desired to be gained by the students through their program at the time of their graduation.

Chapter 2
Emerging Technologies and Educational Requirements in Engineering Education for the Fourth Industrial Revolution

Issa Alghatrifi
Nizwa College of Technology, Oman

Ali S. Al Musawi
https://orcid.org/0000-0002-6893-3216
Sultan Qaboos University, Oman

ABSTRACT

Higher education institutions and their engineering departments have a vital role in fulfilling the new requirements and opportunities of the information and communication technologies (ICT). Therefore, understanding the guidelines to adapt to the new ICT innovations in relation with the Fourth Industrial Revolution such as the internet of things, cloud computing, virtual reality, and artificial intelligence is vital to determine the emerging patterns in their development, delivery, implementation, and assessment. This study aims to define the new educational requirements in engineering education based on the developments of the Fourth Industrial Revolution. Conducting an in-depth analysis of the current literature revealed that although these emerging technologies have been widely used, there are some challenges being faced for their effective use in engineering education. Therefore, the authors provide some guidelines and discuss possible research directions for the use of these technologies in near future.

DOI: 10.4018/978-1-7998-2562-3.ch002

INTRODUCTION

The use of Information and Communication Technologies (ICT) and other emergent technologies have steadily increased in the last few years. This increase led to the emergence of the Fourth Industrial Revolution (FIR), which includes the use of Internet of Thing (IoT), robots, Artificial Intelligence (AI), c-Cloud Computing and Virtual Reality (VR). For example, Cisco experts predict that within the next ten years, IoT devices will reach more than 50 billion at a rate of six devices per person (Cisco, 2013). What distinguishes these devices is the ability to identify, monitor and control every device. Therefore, public and private academic institutions must develop their work settings, keep up their sustainability and efficiency with these advances, and determine the emerging patterns in the delivery of new education. The effect on the engineering education sector is expected to be enormous due to the market demand and curriculum restructure to prepare the next generation students who should deal efficiently with the new emerging technologies.

The main objective of this chapter is to explore the current state of emerging technologies and their educational requirements in engineering education in relation to the FIR developments. It comprises of systematic and detailed review of the current literature in terms of the use of emergent technologies in engineering education, cases and experiments of such technologies in daily educational practices .It also includes the guidelines for the adaptation and the new educational requirements in engineering education.

METHODOLOGY OF THE LITERATURE ANALYSIS

In this section, we report the results of our meta-analysis study, through which we achieve the research objective, and elaborate on the interesting results emerged from the extracted data.

The in Depth Review Approach

The in depth analysis of the literature was conducted in this research to:

1. Provide an overview of the influence of emerging technologies on engineering education.
2. Develop an understanding of these technologies that allow students, teachers and institutions to assess and improve engineering education.
3. Develop an approach that can make these emerging technologies to develop engineering education.

4. Provide recommendations for the use of emerging technologies in higher education engineering by demonstrating the relevance and guidelines for "easy" implementation of emerging technologies of engineering education.

However, findings cannot be generalized due to the limited number of resources, references and studies found in the area under study. This is attributed to the fact that the study area is new and, therefore the accuracy of the findings is not conclusive. Further, inconsistency across the reviewed studies is affected by differences between study methods and processes.

The Review Phases

The in depth analysis of the literature was conducted. The studies were selected using the integrative review process consisted of the following phases (Minch, 2018, Rucker, 2016):

Phase1 The study objective was determined carefully to define the new educational requirements in engineering education based on the FIR's developments. Three guiding research questions were identified as follow:

1. What are the important emerging technologies in relation to FIR?
2. What are the main issues relevant to these technologies in the field of engineering education?
3. How can these technologies work to develop engineering education?

Phase2 The authors search for the relevant literature and select studies and articles using online databases and repositories. References at the end of the chapter list the final resources. The following inclusion criteria were considered:

1. **Quality:** Only studies from globally recognized publications were selected;
2. **Novelty:** All the selected studies are posted in 2010 publications to ensure its newness and originality.
3. **Focus:** Studies related to emerging technologies area, FIR and the studies which are linked to engineering education were selected.

Phase3 Data that answered the research questions were isolated from the selected publications. During this step, findings in relation to the important technologies specifically used in engineering education were located focusing on those found significant by the previous researchers.

Phase4 Data derived from the included studies were assessed rigorously. Data were classified into categories and sub-categories for analysis purposes. Ideas, concepts and thoughts were then critically analyzed, interpreted and organized.

Phase5 Findings of the literature review are discussed and presented below through comparing, augmenting, and through the directions for future trends. Data are presented using tables and figures.

RESULTS

The Current State of Engineering Education

As pointed out in the introduction, the engineering education should take more attentive and critical analysis, in both sides of the engineering equation (education and utilization). This implies that higher education, in general and technical aspects is particularly responsible for preparing the next generation of learners.

Studying the current state of engineering education in global scale is crucial as it analyzes data, capacity and enrolment, jobs and skills, and engineering education trends of the future and gives a holistic view to experts, to cover several gap in engineering education areas. In developed countries, engineering education plays a vital role in knowledge-driven societies adopting different trade and investment strategies (Ruth, 2017). Ruth (2018) identify 81 universities from 22 countries as current global leaders in engineering education, citing 10 of them as a 'current leaders'. Figure 1 presents these institutions.

Figure 1. The 10 institutions most frequently identified as 'current leaders' in engineering education
Source: (Ruth, 2018)

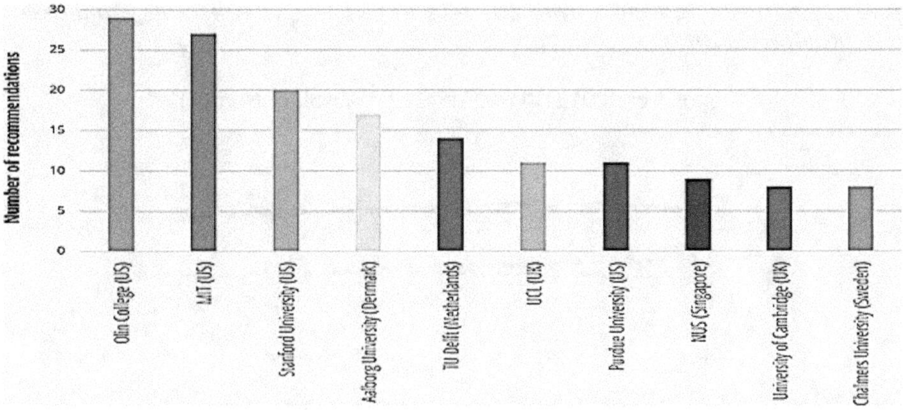

In the global context, Fig 2 highlights the regional 'emerging leaders' from across the world. It shows that countries with propelled engagement in engineering education are located in the southern hemisphere of the globe and the western ones are losing their leadership status.

Figure 2. Global patterns of the 'emerging leaders' in engineering education
Source: (Ruth, 2018)

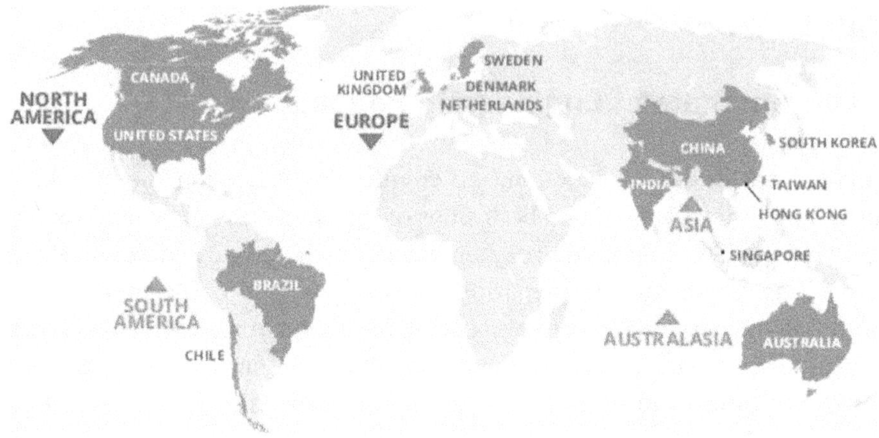

Based on the previous studies there are several issues that affect engineering education, for example the employability of graduating students, capabilities of faculty, and the quality of curriculum design in line with the FIR. Therefore, these issues require immediate attention to improve the enrolment in various disciplines in engineering education. Figure 3 and 4 show the trends and a further analysis of engineering data of the past five years in India.

Figure 3. Engineering trends from 2013-14 to 2017-18 (in absolute numbers)
Source: (Reddy et al., 2018)

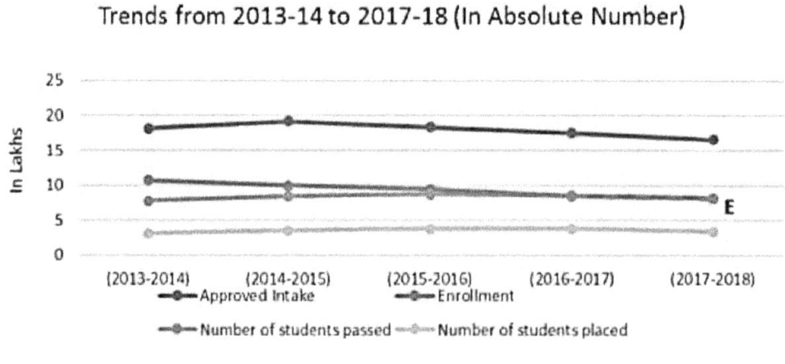

Figure 4. Engineering trends from 2013-14 to 2017-18 in percentages
Source: (Reddy et al., 2018)

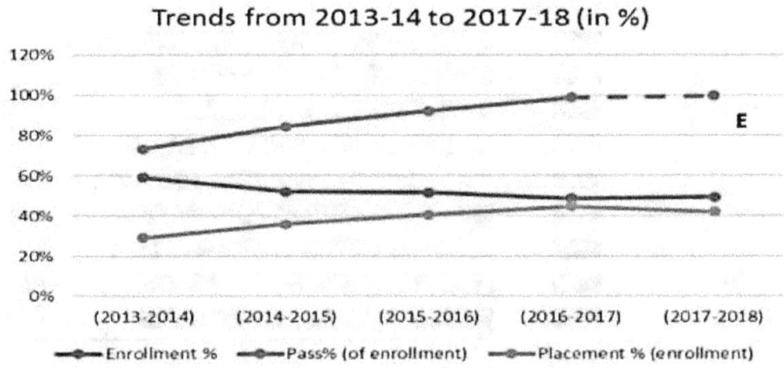

Above data, give several insights that requires more attention from the stakeholders of engineering education. Fig 3 shows that there is a slight increase in the approved intake from 2013 to 2018 and slight decrease in the enrolment from 29% to 42% from 2013-14 to 2017-18. In the same context, and with the students enrolling in the engineering courses we can note that the percentage of students enrolling in the engineering courses indicated a sharp drop in enrollment from 59% in 2013-14 to 49% in 2017-18. In addition, the percentage of students who completed the course in 2016-17 reached 100% compared to 73% in 2013-14. This led to widen the gap between the supply and demand of engineering seats. Moreover, percentages in Fig 4 show modest to progressive increase in students' enrolment. The percentage of enrollment in engineering was 90% in 2017-18, compared to 50% in 2013-14, where the number of seats offered was 15,82,193 compared to 17,36,149 in 2013-

Figure 5. Trends by engineering discipline from 2013-14 to 2017-18
Source: (Reddy et al., 2018)

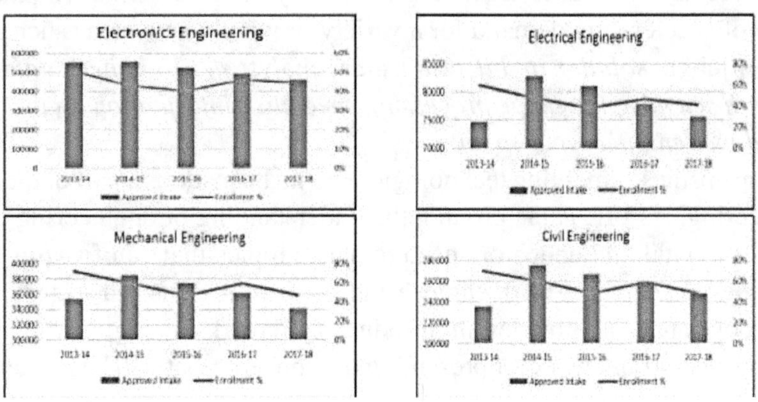

Figure 6. Core skill required for future jobs
Source: (WEF, 2018)

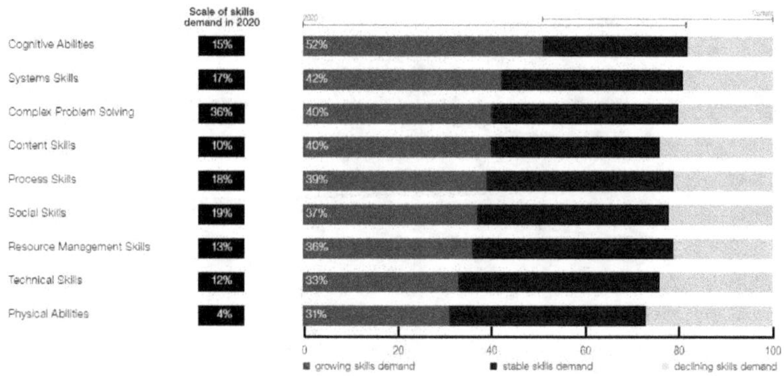

14 (Reddy et al., 2018). In the same context, the analysis by different discipline in engineering education shows that 10 engineering disciplines constitute a large turnout in the number of enrolled students namely, *Civil Engineering, Mechanical, Electrical, and Electronics engineering, Computer Science Engineering, Aerospace Engineering, Mechatronics, and Biomedical Engineering,* Fig 5 shows the trends by engineering discipline in India from 2013-14 to 2017-18.

Moreover, the World Economic Forum (2018) report explained that the future jobs' skill requirements by 2020 would change to involve complex problem solving as one of core skills to deal with the FIR. Fig 6 presents the change in demand for core work-related skills 2015-2020. It is noted that the complex problem solving skills represent 36% of jobs' core skill.

The reviewed literature identified Artificial Intelligence (AI), Internet of Things (IoT), Big Data Analytics, Robotic Process Automation (RPA), Cloud Computing, 3D Printing, Virtual Reality and Block chain as technologies that can be created a huge wave of transformation across industries in the coming years and also identified the skills required for these technologies (Reddy et al., 2018). It emphasizes the evidence of accelerating demand for a variety of wholly new areas such as AI can create new jobs as *solution architect, business analyst, data architect, data scientist, AI research scientist, language processing specialist, information security analyst, and developer engineer and so on.*

To sum up, these emerging technologies should be made as a part of the curricula of engineering. As FIR gains ground, the understanding of engineering education disciplines is vital and educators need to shape topics that can fit with emerging technologies to keep pace with engineering streams in industrial sectors.

The above analysis provides an introductory remarks to engineering education and establishes a framework for presenting the concepts that will be presented in the rest of this chapter by identifying the different types of emerging technologies and their educational applications. Further, it paves the way to discuss the value of these technologies in educational environments. It also helps to analyze the theoretical evidence that the current data provides and the concluding guidelines that should be followed to activate the role of these applications in engineering education and its curricula. Therefore, the next sections should provide the definitions of the most important emerging technologies and to discuss them in details. They explain their importance in engineering education in terms of instructional and utilization parts.

Emerging Technologies in Engineering Education

Internet of Things (IoT)

The Internet of Things term refers to the various devices in our lives to collect and share data. Berhad, (2015) defined IoT as "intelligent interactivity between human beings and things to exchange information and knowledge for a new value creation". It encompasses three main technology components namely; connected things with embedded sensors, connectivity and infrastructure, and analytics and applications (see Fig 7).

Figure 7. Components of IoT
Source (Berhad, 2015)

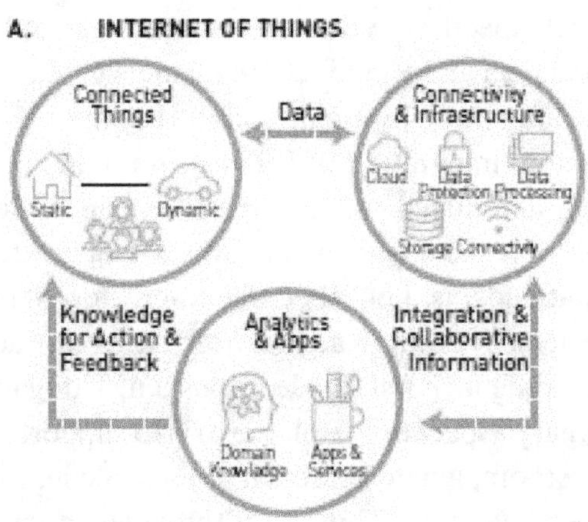

The IoT stands as a core technology in the emerging technology wave, by building a global network that supported new services within information society (Patel & Cassou, 2015). Cisco experts predict that within the next ten years, IoT devices will reach more than 50 billion at a rate of six devices per person. Internet-connected IoT

Figure 8. IoT Road map
Source (Kaukalias and Chatzimisios, 2015)

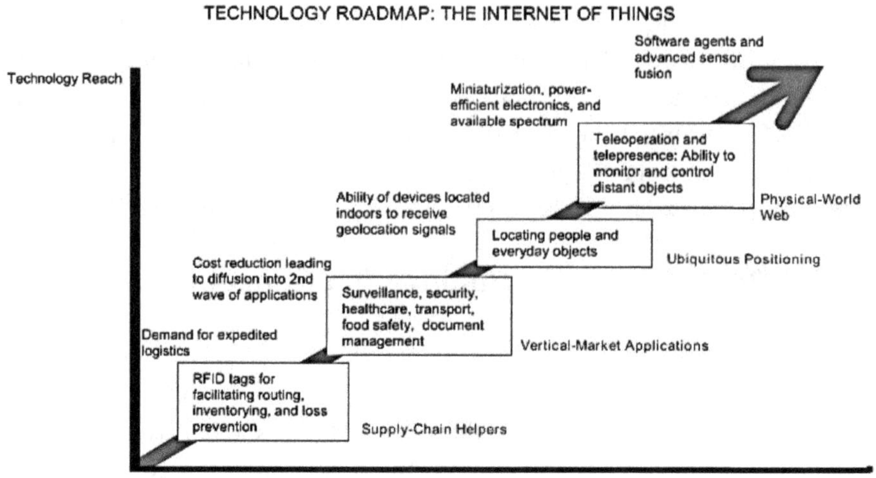

works (Nordrum, 2016; Cisco, 2013) are there to develop a holistic infrastructure for an information society (ITU, 2012). Fig 8 shows that IoT can be classified as the fourth generation of computer development (Abowd, 2016).

Systems approach is one of the emerging solutions that works for integration and interoperability of technology platforms. As observed by the authors, the literary evidence proves that the educational sector should transform the traditional way to digital approach specifically in areas of teaching and learning of the 21st Century students (Dosheela and Binod, 2019; Daniels et al., 2018; Marquez, et al., 2016). Based on the study done by Bagheri and Haghigni (2016) comparing new and traditional education, integration of IoT adds a unique proposition in reduced cost of admin through automated processes; and enhanced learning by increased personalized collaboration and engagement. Gros & López (2016) confirmed these findings explaining that students should become co-creators of knowledge as opposed to being passive learners when they are transferred from traditional education to digital learning environment. Först, Dede, Könsgen, Udugama & Zaman (2017) indicated that the primary aspect of IoT platform is to support learning and teaching using the flipped classroom, a new approach to enhance learning.

Further, knowledge of Internet of Things (IoT) provides an opportunity to explore the careers in aspects of engineering beyond the core curriculum. As a result, universities and institutes have taken steps to adapt IoT to their course and curricular structures (Raikar et al., 2018). Students must be trained in IoT related engineering practice and innovation upheld with solid theoretical foundation. Therefore, establishing an IoT Engineering Laboratory jointly with number of its applications is necessary to carry out basic experiments in various industry applications including *generating information (sensor); acquisition (parts, devices); sharing (Clyde); utilization (Big Data); and application software*. The training of IoT engineers requires different platforms that include operation and training (Kang et al., 2015).

Artificial Intelligence

The term of Artificial intelligence (AI) was first introduced at a conference at Dartmouth College in 1956, in an abstract, theoretical sense only. In the last decade, we began to use it in human life as a consumer technology. AI combines both software and hardware components, for example, the robots on a car assembly line and the software that controls them. The machine learning is the most exciting applications of AI in engineering filed. It depends upon the data generation and analysis. The program of AI equipped with complex algorithms is used to identify mistakes, formulate solutions, and to refine continually. (Daniels et al., 2018).

The impact of AI on engineering job and automation is one of the important issues. There are widespread fears and anxieties surrounding automation, related to replacing humans in certain jobs. However, the authors agree that the long-term benefits of automation can be used to create new opportunities which should be embraced to increase our ability to tackle the largest issues faced by the nations of the earth.

The in-depth development of AI will be bound to accelerate the process of building an individual who, beyond knowledge, is also endowed with critical thinking associated with technical skills and connected to reality. Therefore, we need to provide a professional framework similar to the real world for new engineering education. In this framework the technical simulator expands with professional activities simulator. The major advantage of such a simulator is the integration of the knowledge with skills (Isoc and Surubaru, 2018).

To conclude, it seems that AI technologies make changes in educational landscape. Teaching methods could be changed from personalized learning to educational robots. Four categories should be considered in designing the new engineering education including *customize content, innovative teaching methods, technology enhanced assessment, communication between student and lecturer* (Chassignol et al. 2018).

Cloud Computing

Cloud computing (CC) has virtually unlimited capabilities in terms of storage and processing power. It is an advanced technology having great orientation to resolve all issues related to IoT. That will be made possible by extending its scope to deal with real world things in a more distributed and dynamic manner, for delivering new services in a large number of real life scenarios. (Villari et al., 2016; Botta & de Donato, 2014; Zhou et al., 2013).

CC applies a utility model to produce and consume computing resources, in which the Cloud abstracts all types of computing resources, including storage, as services (i.e. Cloud services). The Cloud user (either application developer or application consumer) can access the Cloud services over the Internet, and the Cloud users pay only for time and services they need. The Cloud can also be scaled to support large numbers of service requests. Ultimately, CC takes care of the micro-lifecycle management of applications, and allows application managers to focus on application development and monitoring. The CC platform is designed to consist of a variety of services for developing, testing, running, deploying, and maintaining applications on the Cloud. Examples of CC platforms are The Amazon® Web Services, Google® App Engine and Microsoft's Windows® Azure platform (Botta & de Donato, 2014). Thus, a novel IT paradigm is formed in which the two complementary technologies like Cloud and IoT merged together to disrupt both Current and Future Internet. We call this new paradigm Cloud IoT (Zhou et al., 2013).

Nevertheless, the application developer needs to consider the running operating systems, networks, load balancing, routers, firewalls, and storage, allowing them to interact with the system by taking into consideration the scalability (Zhou et al., 2013). The contributions of researchers from both the academic and industrial communities should work on the integration of CC and IoT environments. This is to enhance motivation for the continuation of the research by integrating CC and IoT environments for the introduction of new knowledge in this filed.

In engineering education, integration of CC and the Internet of Things represent the next generation of the future Internet. Application of this integration, which is called Cloud IoT, will create a valuable step in new paths for business and research. This integration can draw near future smart cities, automotive, smart home, smart metering and so on. However, all the institutions need to deal with Cloud platforms to be supported with the rapid creation of applications and development, by providing environments and seamless execution of applications, harnessing capabilities of multiple dynamic and heterogeneous resources, to meet the requirements of consumer and industrial sectors (Gubbi et al., 2012).

Virtual reality

One of the new ways to transfer education from traditional way to digital way is through the usage of Virtual Reality (VR). This emerging technology enables dynamic learning by creating virtual environment and also by reflecting the actual environment for learner's interaction. Thus it allows to experiment in a safer environment before the actual application in reality (Dávidekova et al., 2017). In the same context, the education comes into existence by using Augmented Reality (AR) that stimulates the physical reality in a virtual environment side by side with Virtual Reality (VR). This is more attractive and beneficial to learners. Data treatment and numerical simulations with the automated algorithms will be the future of VR in scientific research especially those related to engineering education. In the same context, the VR/AR covers a multi area of industrial engineering systems such as Design of industrial systems, Product design, Process design Logistics, Management of production systems, Maintenance, Safety production systems and so on (Damiani et al., 2018).The automotive industry with the development of Virtual Reality (VR) and Augmented Reality (AR) can provide innovative solutions to several areas of the automotive field from car design to test driving, showroom experience and employee training (Faieza, Azreen & Rohidatun, 2015). In terms of education, Virtual Reality offers both opportunities and challenges for the educational sector. One of the most difficult challenges in education sector is the cost. In recent years, the development of computer hardware and software made feasible to incorporate of Virtual Reality technology in education and into future teaching strategies.

Relevance of Emerging Technologies in Engineering Education

Education Needs

The emerging technologies will affect education in more depth, because of the integration of these technologies in our daily life. Therefore, we need to accept these emerging technologies as new tools for learning and to study the educational needs. Engineering education has been changed by increasing demand for new skills, and from this point of view, we need to start exploration about what kind of technologies that we need to apply in the new education, and what their contribution might be for education (De Vries, Klaasse & Kamp, 2017). The understanding of education needs to study the perceived value of applying these new technologies, especially in relation to stakeholders (student - teacher - organization). Therefore, engineering education needs to be more interactive with technology in teaching and learning (Klopfer, 2016). Based on previous studies, new technologies have been developed and have been widely adopted as educational needs, such as: 3D printing, Internet of Things, BYOD (Bring Your Own Device), Open Source, Learning Management Systems and Virtual Reality (Virtual labs) (NMC, 2016; Veletsianos, 2016). In addition, there are open tools for self-learning (social media) like YouTube, Google search, Twitter, Power point, Google docs, Facebook, Skype, etc., so that it cannot be denied its importance as educational needs.

Perceived Values

Before studying the perceived values for emergent technologies, we need to discover three major values in engineering education as interrelated stakeholders decisive for the usability of technology (student – teacher and the organization) to answer the question: what is the perceived value for the students; teacher and what are the results perceived values for the organization? (De Vries Klaasse & Kamp, 2017). From the student side the higher education institutions should be poised to prepare students effectively for the labour market by providing necessary skills for more innovative and productive results (Kamp, 2016; World Economic Forum, 2016 & 2018; Klopfer, 2016). Mesquita (2016 indicated that the success technology implementation in any education relates to several critical individual behaviour that related to the instructor such as support, training, availability of equipment and applications.(Jaschik & Lederman (2015) indicated that the some instructors are sceptical about the value of technology, but with availability of support and training, the appreciation seems to increase. Lastly, the existence of a specific and planned strategy in introducing emerging technologies in the enterprise creates benefit for the success of the organization and all of that play a major role to avoid constrainers like intellectual property, credits, development of new materials, etc. In addition, it will create a perceived value of the organizations by partnering with industry to enable and equip students with the latest skills for the staff development. It will also equip instructors with the skills about the usage of technology. In fact it enables the faculty to update the ways in which they deliver content and assess student learning (Becker et al., 2017; Briggs et al, 2016; Janssen et al., 2016; Jaschik & Lederman, 2015; NMC, 2016; King & South, 2017).

DISCUSSION OF DATA

Cases and Experiments of Such Technologies in Daily Educational Practices

The review of some experiment by using emerging technologies and the presentation of some case studies is of critical importance. It is used to emphasize the relevance of applying these technologies and their benefits in education.

The writers focus here on Virtual Reality stress the great impact on teaching and learning more than any other emerging technology. Moreover, it is widely used at reasonable costs and is available on the Internet, making it easily accessible to both students and researchers. In addition, its effectiveness in engineering education studies is well known compared to other more recent technologies.

Virtual Reality (VR) is a "computer technology that uses special headsets to generate images, sounds and other sensations that imitate a real environment to simulate a user's physical presence in 3D 360- degree environment" (De Vries, Klaasse & Kamp, 2017). The famous tool in VR education environment is VR headset: head-mounted goggles with a screen in front of the eyes. Fig 9 presents this tool.

Figure 9. Virtual Reality Headset (Goggles)
Source: (www.amazon.com)

Usage of this technology provides five educational benefits that can be classified as a benefit to education as follows: spatial knowledge representation, experiential learning, engagement, contextual learning and collaborative learning (De Vries, Klaasse & Kamp, 2017). During 2016-2017, the use of a smartphone for VR experience has enlarged the opportunity for education. It allows experimentation more than using Virtual Reality Headset (Goggles). During the same period, a virtual engineering lab was erected. This lab provides more realistic experiences where students can hold, rotate and fit together with virtual parts as they would with their real hands, Pennsylvania State University is a leader in this field (Wertz, 2016). In engineering education, the VR is a helpful tool for subjects such as engineering and architecture. It can be used to design and manipulate virtual structures with the perception of the meaning of 'value'.

Guidelines to Adapt Emergent Technologies in Engineering Education

To develop and adapt the emergent technologies into their instructional methods, engineering educators, instructors, and curriculum designers need to apply and follow guidelines in the development, delivery, implementation, and evaluation stages. These are discussed below (Al Musawi, 2011; Riojas, Lysecky, Rozenblit, 2012):

Development Guidelines

1. Analyze the readiness of their educational institutions by checking the availability of hardware and software, administrative and financial support as well as copyright procedures, and course delivery mechanisms.
2. Consider the needs of their learners to be assessed in the areas of their thinking levels, scientific backgrounds and performance levels.
3. Find out ethical issues involved in the instructional delivery such as diversity, internet accessibility for the special needs students, and plagiarism.
4. Analyze the curriculum content to identify the main concept, theories, facts, and ideas.
5. Analyze the instructional goals and determine the learning objectives using the Bloom Taxonomy-Cognitive Domain and behavioral descriptions of the learning outcomes (Riojas, Lysecky, Rozenblit, 2012).

Delivery Guidelines

1. Select appropriate instructional modes such as demonstration, simulations, modelling, field trips, and case studies.
2. Plan for technologies to enable the implementation of the inductive teaching methods of problem/inquiry and project-based sessions. Software-based technologies cover a large number of engineering related topics. Hardware-based technologies should be related to computer, electrical and mechanical engineering (Riojas, Lysecky, Rozenblit, 2012).
3. Build the teaching-learning environment using technologies that may comprise wireless technologies, electronic games, assistive technologies, intelligent robots in addition to the computers and laptops. Software/applications technologies, multimedia and social media networks can be utilized to explain engineering concepts that need imagination such as those at micro-and-macro levels.
4. Design the interface for online course/content; and test its user-ability, accessibility, navigation, and interactivity.
5. Prepare to provide the students with different traditional and electronic resources such as private tuition, email/online help, books, AV library, and learning management platforms.

Implementation Guidelines

1. Use the traditional and synchronous/asynchronous online instruction.
2. Provide the course content and resources ensuring the learner interaction and outcome achievement.

3. Focus on learners' engagement by using collaboration, activities, teamwork and threaded discussions.

Assessment Guidelines

1. Evaluate the students' achievement of learners' performance.
2. Assess user-ability, usability of the education process and its effectiveness.

The Educational Requirements in Engineering Education

Responding to Market Demand

At the end of 2016, the World Economic Forum announced tipping points expected to occur in market demand by 2025. Based on the FIR and future market, the statistics show these tipping points (see Fig 10).

Figure 10. Tipping points expected in the market to occur by 2025.
Source: (WEF, 2015)

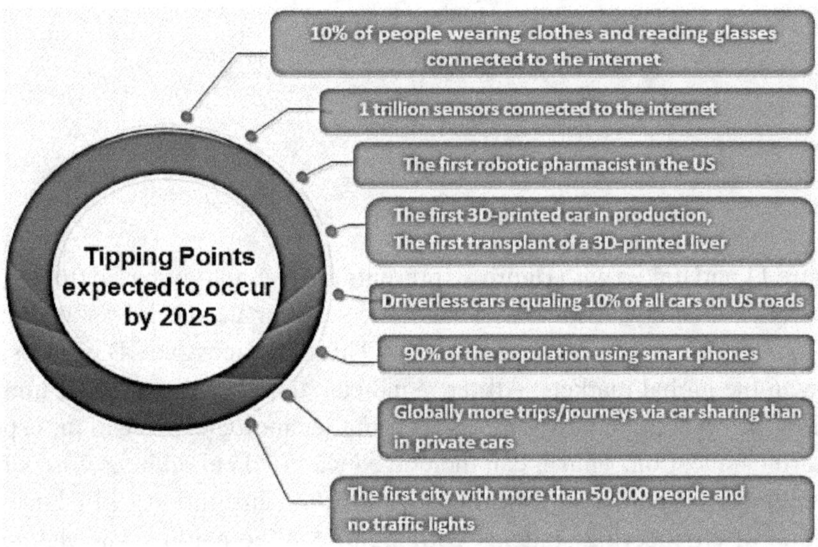

The phenomenal scientific development of IoT, AI, big data and the growth of the sharing economy will lead to transform the world in the upcoming years. From the figure above and starting from the daily tasks such as the wearing clothes to driving habits, we can understand that emergent technologies will drive most of our daily life tasks. The industry 4.0 is in our markets and we all should get ready for the changes as this pushes experts and curriculum designers to redesign the educational curricula so that the coming generation will be able to control this emergent technologies revolution and be on line with it, starting from Internet of Things to Google glasses. For example, as mentioned in the figure above the 3D printers are now present in the local markets to produce some simple materials, but looking ahead to the future, we need to put in our curriculum design that the architects should be able to control this emerging technologies align with the pace of alteration, for example in designing future buildings and various other products.

In addition, Fig 11 presents the technology-based tipping points for the future society:

Figure 11. Technology based tipping points for the Future Society.
Source (WEF, 2015)

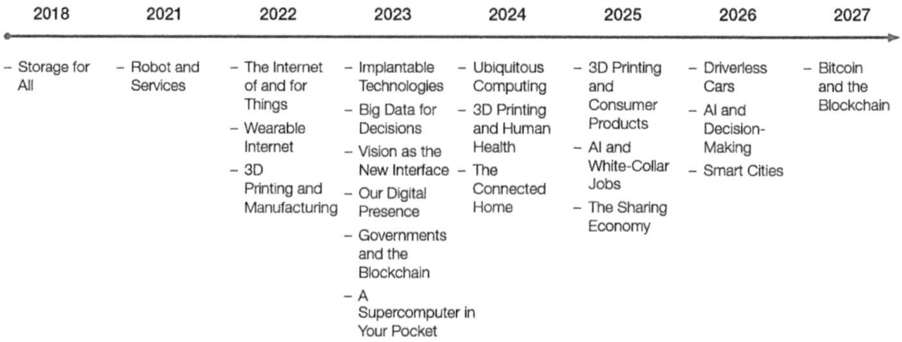

Figure 11 and linked with figure 10 presents the technology-based tipping points for the future Society. It indicates that robots dominate the ladder of emerging technologies. This is followed by Internet of Things devices and 3D printers, which are now in the global markets. All these indicate that there must be an immediate shift from the existing education, to emerging technologies education, especially engineering education. Thus it can include education on ubiquitous computing.

In engineering education, the higher education institutions will have a new generation of engineering students with significant computing knowledge. They expect that academic institutes will introduce them to emerging technologies to meet the demands of the market. However, the academic institutions need to adopt appropriate strategies to meet the innovative educational demand (Martin, 2017). One of the most important tools to face these demands is curriculum restructure.

Curriculum Restructure

Developing a modern education curriculum requires a sustainable design showing new technologies and processes in an integrated manner. Under FIR, involving instructors and students directly in developing the education curriculum is crucial for learning. (De Vries, Klaasse & Kamp, 2017).

In the past, Engineering Education (represented by Engineering Education 1.0) lacked interactivity tools and this, with increasingly use of mobile phones and sharing of content virtually, has begun the era of writing content and sharing ideas reaching what is called "Engineering Education 2.0. Shortly generation 2.0 thereafter was called "Engineering Education 3.0" by the use Google Glass and Virtual Reality headset (Goggles). Due to recent advances in technology and the emergence of active collaboration through new technologies such as Internet of Thing (IoT), robots, Artificial Intelligence (AI), Cloud Computing and Virtual Reality (VR), Engineering Education 4.0 appeared, intelligence has become embedded in devices. These "smart" devices have the ability to interact with humans and other smart devices.

IoT applications, Virtual Reality (VR) and Augmented Reality (AR) have been based on fragmented software implementations for specific systems and cases, and this applies to Engineering Education in schools and universities. However, from this perspective, we need to be ready for change and need to restructure curriculum to keep pace with the change in the forthcoming Engineering Education 4.0. One of the aims of this chapter is to look at what has been discussed in the preparation of educational curricula related to modern education, especially with regard to Engineering Education 4.0 and choose the best strategy for successful adoption of new emerging technology in higher education.

It seems difficult for any university or college on a global scale to get to the point of transformation in Engineering Education 4.0 without concrete understanding of what is the core concepts of this new generation of engineering education. We need to understand the readiness of Higher Education Institution (HEIs) and ICT personnel for migration to its adoption. Based on the study conducted by Prasad et al. (2018) and Triki (2016) the development of curricula should stress the need for flexibility in restructuring the technical and engineering education programmes to improve the student performance, innovation and self-development for successful implementation of Engineering Education 4.0. That means that effective curriculum structure should be based on sound instructional design strategies. According to Katharina & Dominic (2015), combining real and virtual worlds' information with the use of virtual resources, for example glasses for Virtual Reality, will lead to enhance education in HEIs. Moreover, Huba & Kozák (2016), Richter et al (2015),

Schuster (2015) pointed out that virtual learning environments (VLEs) will be used for developing knowledge and skills; and teachers with students will meet to achieve this transformation. For instance, Engineering Education of IoT, considered as one of the most significant activities in society at present in scope of the rapid growth of IOT application and devices. It dramatically influenced the educational process in recent decades (Kin & Kam, 2015). Therefore, the technological innovation has now changed the approach of how to successfully integrate the best tools into the teaching process. All of that lead to provide a virtual, shared and intelligent experimental teaching environment, so the students in this environment play an active role rather than the traditional passive role of recipient of information. In teaching the IoT curriculum, the teacher plays the role of guide by providing guidelines and resources to achieve the project goals and support the student activity (Costa Pereira, 2014; Vujović et al, 2014).

It seems that the current studies on Engineering Education 4.0 have utilized a specific technology-adoption model or integrated multiple models to assess its successful implementation in HEIs. This preliminary evaluation should suffice to prepare the stage for Education Engineering 4.0 to be compatible with the new industrial revolution 4.0. Therefore, to identify the requirements of industry 4.0 and teaching engineering curriculum in HEIs landscape, we need to illustrate possible sources to adequate learning scenario for specific learning goals. Based on the in-depth analysis of studies investigating the requirements of industry 4.0 and curriculum in higher education, Fig 12 illustrates the model of restructuring the Engineering Education skill-based curriculum.

From the model of restructuring the engineering education to a skill-based curriculum, we find that there is a three-way interaction between the learning work environment, competences profiles and personal competences. This three-way interaction, will affect characteristics of curriculum restructure for emerging technologies. This leads us to think that research and work should be done to develop a new conceptual framework based on new structures which are compatible with these emerging technologies. We need to rebuild the engineering education by restructuring it to a skill-based curriculum (Lensing and Friedhoff, 2018). We need to add a new element to this model with the name emerging technologies competences which refers to the combination of humans and emergent technologies. This will work side by side to contain new structures which are compatible with these emerging technologies. This competence includes micro-tasking, workflows and ecosystems. Micro-tasking refers to the independence of emerging technology in work or the need for human intervention. Workflows refer to the complete acquisition of workflow and data flow by the emerging technology. Ecosystems refer to the combination of the cognitive processing of the data and machine-based computing to come up with compatible model which combines learning environment, work environment,

Figure 12. Identification of competence requirements for industry 4.0 and IoT curriculum in higher education (adapted from Lensing and Friedhoff, 2018).

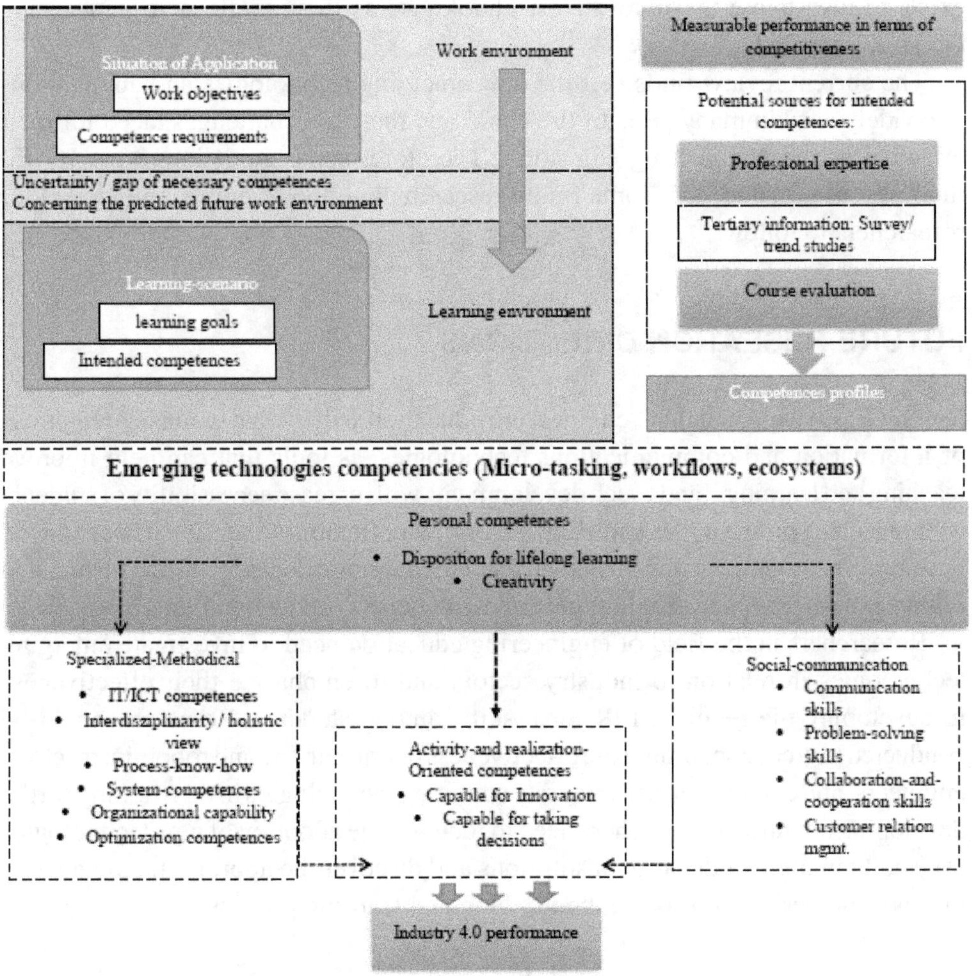

competence profiles, and personal competence. All of them will lead the curriculum designer to design the skills of the student in engineering education that must be acquired and integrated with all competences suggested in this model.

CONCLUSION

The main objective of this chapter is to define the new educational requirements in engineering education based on the FIR's developments. To achieve this objective, the authors conduct an in depth analysis of the current literature. This review is limited to the time when the reviewed literature was published and, thus, a need exists to investigate the innovative technologies as they continue to appear and being investigated.

The current review finds that the new emerging technologies and applications are widely used in many parts of the world and there are challenges faced for their effective use in engineering education. Therefore, some guidelines were set for their best use. In addition, some future research directions were recommended for researchers to follow.

FUTURE RESEARCH DIRECTIONS

Rosen (2009) expected that "engineering education will evolve to make greater use of information and communications technologies, as tools that can help improve teaching and learning." and "such developments will increase accessibility of students to engineering programmes and allow them greater flexibility" (p. 29). These visions have become true as we nowadays see that engineering educators utilize virtual and online experiments and simulations to ease concepts' understanding.

Researchers in the field of engineering education need to investigate emerging technologies in relation to industry sectors and to emphasize their effectiveness in developing the required FIR jobs' skills among students. Research should be conducted in areas concerning the effectiveness of training IoT and robots instructors, engineers, and explore ways to integrate emerging technologies in this training. In the developing countries, researchers need to focus on the adoption of new technologies especially in higher education institutions and discover the appropriate methods to take advantages of their use in the teaching and learning processes.

REFERENCES

Abowd, G. D. (n.d.). Beyond Weiser: From Ubiquitous to Collective. Computing. *IEEE Computer, 49*(1), 17–23.

Al Musawi, A. (2011). Blended Learning. *Journal of Turkish Science Education, 8*(2), 3-8.

Bagheri, M., & Haghighi, M. (2016). The Effect of the Internet of Things (IoT) on Education Business Model. In *2016 12th International Conference on Signal-Image Technology & Internet-Based Systems (SITIS)*. IEEE Computer Society.

Becker, A., Cummins, M., Davis, A., Freeman, A., Hall Giesinger, C., & Ananthanarayanan, V. (2017). *NMC Horizon Report: 2017 Higher Education Edition*. Austin, TX: The New Media Consortium.

Berhad, M. (2015). *National Internet of Things (IoT) Strategic Roadmap*. Available: http://www.mimos.my/IoT/National_IoT_Strategic_Roadmap_Summary.pdf

Botta, A., & de Donato, W. (2014). On the Integration of Cloud Computing and Internet of Things. In *Proceedings of the International Conference on Future Internet of Things and Cloud* (pp. 23 – 30). IEEE Conference Publications. 10.1109/FiCloud.2014.14

Briggs, B., Foutty, J., & Hodgetts, C. (2016). *Tech Trends*. Deloitte University Press.

Chassignol, M., Khoroshavin, A., & Bilyatdinova, A. (2018). Artificial Intelligence trends in education: A narrative overview. *Procedia Computer Science*, *136*, 16–24. doi:10.1016/j.procs.2018.08.233

Cisco. (2013). *Industrial smart solutions: connecting the factory to the enterprise*. Whitepaper. Available: https://www.cisco.com/c/dam/global/en_ph/assets/tomorrow-starts-here/files/white_paper.pdf

Damiani, L., Demartini, M., Guizzi, G., Revetria, R., & Tonelli, F. (2018). Augmented and virtual reality applications in industrial systems: A qualitative review towards the industry 4.0 era. *IFAC-PapersOnLine*, *51*(11), 624–630. doi:10.1016/j.ifacol.2018.08.388

Daniels, J., Sargolzaei, S., Sargolzaei, A., Ahram, T., Laplante, P. A., & Amaba, B. (2018). The internet of things, artificial intelligence, blockchain, and professionalism. *IT Professional, 20*(6), 15-19. doi:10.1109/MITP.2018.2875770

Dávidekova, M., Mjartan, M., & Greguša, A. (2017). Utilization of Virtual Reality in Education of Employees in Slovakia. *The 8th International Conference on Emerging Ubiquitous Systems and Pervasive Networks*. Available: https://pdf.sciencedirectassets.com

De Vries, P., Klaasse, R., & Kamp, A. (2017). Emerging technologies in engineering education: can we make it work? *Proceedings of 13th International CDIO Conference*, 1-12.

Dosheela, D., & Binod, K. (2019). *Exploring the Internet of Things (IoT) in Education: A Review*. Information Systems Design and Intelligent Applications. Available https://link.springer.com/chapter/10.1007%2F978-981-13-3338-5_23

Faieza, A., Azreen, A., & Rohidatun, M. (2015). Virtual Reality Application In Design, Prototype and Test Drive Processes In Automotive Industry. *International Journal of Advance Research In Science and Engineering, 4*(1). Available: https://pdfs.semanticscholar.org/d485/2354c2349bfe2a0aad2919587a8675071dee.pdf

Förster, A., Dede, J., Könsgen, A., Udugama, A., & Zaman, I. (2017). Teaching the Internet of Things. GetMobile. *Mobile Computing and Communications, 20*(3), 24–28.

Gros, B., & López, M. (2016). Students as co-creators of technology-rich learning activities in higher education. *International Journal of Educational Technology in Higher Education, 13*(1), 28. doi:10.118641239-016-0026-x

Gubbi, J., Buyya, R., Marusic, S., & Palaniswamia, M. (2012). *Internet of Things (IoT): A Vision, Architectural Elements, and Future Directions*. Available: https://arxiv.org/ftp/arxiv/papers/1207/1207.0203.pdf

Huba, M., & Kozák, Š. F. (2016). From E-learning to Industry 4.0. *International Conference on Emerging eLearning Technologies and Applications (ICETA)*, 3-108. DOI: 10.1109/ICETA.2016.7802083

Isoc, D., & Surubaru, T. (2018). *Engineering Education Using Professional Activity Simulators*. Available https://link-springer-com.ezproxy.utm.my/chapter/10.1007/978-3-030-11932-4_50

ITU. (2012). *Overview of the Internet of Things*. Retrieved from: http://www.itu.int/ITUT/recommendations/rec.aspx?rec=Y.2060

Janssen, D., Tummel, C., Richert, A., & Isenhardt, I. (2016). Higher Education – Immersion as a Key Construct for Learning 4.0. *International Journal of Advanced Corporate Learning*, *9*(2), 20–26. doi:10.3991/ijac.v9i2.6000

Jaschik, S., & Lederman, D. (2015). *The 2016 inside Higher Ed Survey of College and University Chief Academic Officers*. Washington, DC: Inside Higher Ed and Gallup. Available: https://www.insidehighered.com/booklet/survey-college-and-university-chief-academic-officers

Kamp, A. (2016). Engineering Education in a Rapidly Changing World; Rethinking the Vision for Higher Education (2nd ed.). 4TU-Centre for Engineering Education.

Kang, Y., Han, M., Han, K., & Kim, J. (2015). A Study on the Internet of Things (IoT) Applications. *International Journal of Software Engineering and Its Applications*, *9*(9), 117–126. doi:10.14257/ijseia.2015.9.9.10

Katharina M., Dominic G. (2015). In-Factory Learning-Qualification for the Factory of the Future. *Acta Universitatis Cibiniensis – Technical Series 2015, 66*, 159-164. DOI: doi:10.1515/aucts-2015-0046

Kaukalias, T., & Chatzimisios, P. (2015). *Internet of Things (IoT)*. IGI Global. Available: https://www.irma-international.org/viewtitle/112465/

Kin, S., & Kam, C. (2015). The Impact of the Internet of Things on Engineering Education. Academic Press.

King, J. & South, J. (2017). *Reimagining the Role of Technology in Higher Education: A Supplement to the National Education Technology Plan*. U.S. Department of Education, Office of Educational Technology.

Klopfer, J. (2016). *Augmented Learning: Research and Design of Mobile Education Games*. MIT Press.

Lensing, K., & Friedhoff, J. (2018). Designing a curriculum for the Internet-of-Thing- Laboratory to foster creativity and marker mindset within varying target group. *8th conference on learning factories 2018- Advance Engineering Education and Training for Manufacturing Innovation*. www.sciencedirect.com

Marquez, J., Villanueva, J., Solarte, Z., & Garcia, A. (2016). IoT in education: Integration of objects with virtual academic communities. *New Advances in Information Systems and Technologies*, *1*, 201-212. Available: https://link.springer.com/chapter/10.1007/978-3-319-31232-3_19

Martin, J. (2017). Editorial: Learning Strategies in Engineering Education Using Virtual and Augmented Reality Technologies. *EURASIA Journal of Mathematics Science and Technology Education*, *13*(2), 297–300. Available: http://www.ejmste.com

Minch, C. (2018). How to Write a Literature Review. *Centre for Effective Services*. Available: https://www.teachingcouncil.ie/en/_fileupload/Research/Literature-Review-Webinar.pdf

NMC- The New Media Consortium. (2016). *NMC Horizon Report (2016): Higher Education Edition*. Austin, TX: The New Media Consortium. Available: https://www.sconul.ac.uk/sites/default/files/documents/2016-nmc-horizon-report-he-EN-1.pdf

Nordrum, A. (2016). *Popular Internet of Things Forecast of 50 Billion Devices by 2020 Is Outdated*. Available: https://spectrum.ieee.org

Patel, P., & Cassou, D. (2015). Enabling high-level application development for the Internet of Things. *Journal of Systems and Software*, *103*, 62–84. doi:10.1016/j.jss.2015.01.027

Pereira, C. (2014). *Perspectives and Approaches for the Internet of Things* (Master Thesis). Faculty of Science and Technology, NOVA University Lisbon.

Prasad, J., Goswami, A., Kumbhani, B., Mishra, C., Tyagi, H., Jun, J. H., ... Das, S. K. (2018). Engineering curriculum development based on education theories. *Current Science*, *114*(9), 1829. doi:10.18520/cs/v114/i09/1829-1834

Raikar, M., Desai, P., Vijayalakshmi, M., & Narayankar, P. (2018). *Upsurge of IoT (Internet of Things) in engineering education: A case study*. Available: https://ieeexplore.ieee.org/document/8554546/

Reddy. (2018). *Engineering Education in India – Short & Medium Term Perspectives*. Available https://www.aicte-india.org

Richter, A. (2015). LEARNING 4.0: Virtual Immersive Engineering Education. *International Best Practices and Applications*, *11*, 51–66.

Riojas, M., Lysecky, S., & Rozenblit, J. (2012). Educational Technologies for Precollege Engineering Education. *IEEE Transactions on Learning Technologies*, *5*(1), 20–37. doi:10.1109/TLT.2011.16

Rosen, M. (2009). Engineering Education: Future Trends and Advances. *Proceedings of the 6th WSEAS International Conference on Engineering Education*. Available: https://pdfs.semanticscholar.org/c85b/7aa6297ce266065bceed4d46b4851252bd31.pdf

Rucker, M. (2016). *A Brief Overview of Three Types of Literature Review*. Available: https://unstick.me/brief-overview-three-types-literature-review/

Ruth, G. (2017). *The global state-of-the-art in engineering education Outcomes of Phase 1 benchmarking study*. Available: https://www.cti-commission.fr/wp-content/uploads/2017/10/Phase-1-engineering-education-benchmarking-study-2017.pdf

Ruth, G. (2018). *The global state of the art in engineering education*. School of Engineering Massachusetts Institute of Technology. Available: http://neet.mit.edu/wp-content/uploads/(2018)/03/MIT_NEET_GlobalStateEngineeringEducation(2018).pdf

Schuster, K., Plumanns, L., Groß, K., Vossen, R., Richert, A., & Jeschke, S. (2015). Preparing for Industry 4.0 – Testing Collaborative Virtual Learning Environments with Students and Professional Trainers. *International Journal of Advanced Corporate Learning*, *8*(4), 14. doi:10.3991/ijac.v8i4.4911

Triki, N. (2016). Innovations and Curriculum Development for Technical and Engineering Education in Libya. *Literacy Information and Computer Education Journal*, *7*(4). Available https://infonomics-society.org/licej/. doi:10.20533/licej.2040.2589.2016.0324

Veletsianos, G. (2016). *Emergence and innovation in digital learning: Foundations and applications*. Athabasca University Press. doi:10.15215/aupress/9781771991490.01

Villari, M., Al-Anbuky, A., Celesti, A., & Moessner, M. (2016). Leveraging the Internet of Things: Integration of Sensors and Cloud Computing Systems. *SAGE Journal*. Available on https://journals.sagepub.com/doi/full/10.1177/155014779764287

Vujović, V., Maksimović, M., & Perišić, B. (2014). Collaboration in Software Engineering classroom. *12th IEEE International Conference on Emerging eLearning Technologies and Applications*, 40(6), 505-510.

WEF-World Economic Forum. (2015). *Deep Shift Technology Tipping Points and Societal Impact: Survey Report, September 2015*. Available: http://www3.weforum.org/docs/WEF_GAC15_Technological_Tipping_Points_report_2015.pdf

WEF- World Economic Forum. (2018). *The Future of Jobs Report (2018), Insight Report*. Available: http://www3.weforum.org/docs/WEF_Future_of_Jobs_(2018).pdf

Wertz, P. (2016). *Engineer receives funding to support teaching in a virtual environment*. Available: https://news.psu.edu/

World Economic Forum. (2016). The Future of Jobs Employment, Skills and Workforce Strategy for the Fourth Industrial Revolution. *Global Challenge Insight Report*. Available: http://www3.weforum.org/docs/WEF_Future_of_Jobs.pdf

Zhou, J., Leppanen, T., & Harjula, E. (2013). Cloud Thing: A Common Architecture for Integrating the Internet of thing with Cloud Computing. In *CSCWD, 2013*. IEEE.

ADDITIONAL READING

Lomask, M., Crismond, D., & Hacker, M. (2018). Using Teaching Portfolios to Revise Curriculum and Explore Instructional Practices of Technology and Engineering Education Teachers. *Journal of Technology Education*, *29*(2), 54–72. doi:10.21061/jte.v29i2.a.4

Riojas, M., Lysecky, S., & Rozenblit, J. (2012). Educational Technologies for Precollege Engineering Education. *IEEE Transactions on Learning Technologies*, *5*(1), 20–37. doi:10.1109/TLT.2011.16

Veletianos, G. (2016). *Defining characteristics of emerging technologies and emerging practices. Publisher* (pp. 3–16). Athabasca University Press.

Weil, V. (2003). Ethics across the Curriculum: Preparing Engineering and Science Faculty to Introduce Ethics into Their Teaching, Proceedings of National Academy of Engineering- National Academies Workshop on Emerging Technologies and Ethical Issues in Engineering, October 14-15, 2003.

KEY TERMS AND DEFINITIONS

Artificial Intelligence: Is a combination of both software and hardware components, for example, the robots on a car assembly line and the software that controls them.

Cloud Computing: Is "the practice of using a network of remote servers hosted on the Internet to store, manage, and process data, rather than a local server or a personal computer" (http://english.oxforddictionaries.com/cloud%20computing).

Internet of Things: Is an intelligent interactivity between human and things to exchange information and knowledge for new value creation. It encompasses three main technology components, namely connected things with embedded sensors, connectivity and infrastructure, and analytics and applications.

Virtual Reality: Is "an artificial environment that is created with software and presented to the user in such a way that the user suspends belief and accepts it as a real environment" (https://whatis.techtarget.com/definition/virtual-reality).

Chapter 3
Designing a Training Platform for Higher Education Engineering Instructors in the Digital Era

Fırat Sarsar
https://orcid.org/0000-0002-3611-8137
Ege University, Turkey

Özge Andiç Çakır
Ege University, Turkey

ABSTRACT

Higher education (HE) should focus on solving the following critical educational problems: (1) using technology and (2) fostering education by new creative learning techniques. In this chapter, the authors indirectly talk about using new technologies in education. There are many reasons that make this choice challenging such as believing in the benefits, having enough knowledge, accessing alternative technological sources, etc. To facilitate this, they introduce an online learning platform for engineering instructors in HE. Moreover, according to their experiences in the field of education and engineering, instructors in HE should improve and revise their skills and knowledge. It is important to enhance knowledge on content, technology, and pedagogy; therefore, this training platform itself focuses on improving those skills necessary for instructors in HE for maintaining an effective learning process. This chapter mainly focuses on creating a course for higher education engineering instructors and a TERR model that is proposed by the authors.

DOI: 10.4018/978-1-7998-2562-3.ch003

INTRODUCTION

There has been a global shift towards new engineering trends arising from the need for environmentally responsible products, sustainability as a design approach, fast-changing technology that boosts innovation and seeking solutions by interdisciplinary vision. These bring in not only industrial but also societal growth directing the focus on engineers with new skills and competences. To adopt the needs of labour market, today's engineers need to improve their soft skills, such as communication techniques, ethics, environmental conscience, social media, etc. to confront socio-scientific issues described here. Another challenge is the demand for a qualified workforce for technology and innovation. New ways of (student-centred) learning are characterised by personalisation, engagement, use of digital media, collaboration and bottom-up practices. Thus, the methods based on which the learner is the creator of learning content are emerging and are facilitated by the exponential growth in open education resources available via the internet.

Today's improvements in technology provide materials, tools and techniques that make life easier and enable us to work more productively. The devices such as smart boards, classroom PCs, projectors and digital information boards have been introduced to daily classroom activities decades ago. Today, nearly every classroom has such infrastructure, nearly every student uses a smart mobile device but still, a low number of classroom activities are centred on interactive/visual education in classes. Information and communication technologies (ICT) enable the learners to interact within a learning environment and improve the quality of learning experience for students. Nevertheless, EU Modernisation agenda for higher education focuses on as Ossiannilsson (2018) mentioned that promoting to put student in the centre of teaching and learning through greater EU support for meaningful teaching, curriculum design and use of ICT effectively.

Looking from the students' perspective, they are generally excited for using ICT as learning materials because they are familiar with using ICT in their daily life for information searching and communicating. On the contrary, not born into an ICT supported world like their students, higher education instructors have awareness of technology, are highly knowledgeable in their area, and thus, they might not be expected to be resistant to progress and change in the methods of teaching.

It can be emphasized that successful implementation of instructional technologies in higher education is possible by focusing on the teachers' beliefs, approaches and their expectations. Disciplines and institutional cultures were among the factors influencing the teachers' attitudes. Analysing the interviews with a group of higher education teachers they observed within a 10 years period, Englund et al (2016) concluded that there is a significant difference between the conceptions of experienced teachers vs non-experienced ones. Thus, a conceptual change should be supported with professional development activities.

This chapter includes a case study driven from an ERASMUS+ project entitled "Student-Centred Learning Approaches (ESCOLA)". The project aims to develop a digital learning environment to improve engineering instructors' and students' ICT knowledge by using digital tools and resources. Within the scope of the project, an online learning platform for teachers introducing the use of new technologies was developed and pilot tests were carried out and the opinions of members of the engineering faculty were examined. Analysis, Design, Develop, Implement, and Evaluate (ADDIE) model was employed for online learning platform design. Each step of ADDIE model is explained and detailed by reflecting experiences of designing process from the Turkish partner's viewpoint of the project. After the course design by ADDIE model, the professional development was demonstrated by the TERR method further explained in this chapter.

ADDIE as an Instructional Design Model

ADDIE is one of the very-well known instructional design models) Educators design instructional process according to the stages of the ADDIE model. In the analysis stage, they determine the needs of the learners about the instructional process. Also, this stage includes creating instructional targets and figuring out what should be educated to achieve the instructional objectives. In the design stage, educators plan how to reach their goals. They should be creative about that plan and they make an expansive review. In the development stage, every part of instruction should be planned practically, and this plan should be enough to fulfil the plan created during the design stage. In the implementation phase, educators should be testing the instructions. They can take advantage of trial project or analogous beta in this stage. The last stage is evaluation stage. In the evaluation stage, educators get feedback about the curriculum and make the fitting changes in accordance with the curriculum (Cheung, 2016). The main purpose of the Analysis is to identify the possible causes of a performance gap (Branch, 2009). The Analysis phase is to identify learners' characteristics, their needs, instructional goals, key skills and so on. Also, the content of the training can be analysed considering course texts, sample curriculum, and websites of the course with a similar focus (Peterson, 2003).

The next step of ADDIE is Design and it means to create the outline of educational strategies. It clarifies the expected learning outcomes, learning activities and assessment tools. Branch (2009, p.17) defines this phase as verifying the desired performances and appropriate testing methods. In the design stage, educators plan how to reach their goals. They should be creative about that plan and they make an extensive review (Cheung, 2016). Park & Wang (2019) used deciding learning destinations and procedure, choosing assessment methods and frequency, preparing online feedback tool and resources, and deciding evaluation methods for their module, in the design stage of their study.

In the Development stage, every part of instruction should be planned practically, and this plan should be enough to fulfil the plan created during the design stage (Cheung, 2016). Park & Wang (2019) used making ready their module content, tools, online resources, materials, and evaluating the method for their module, in the development stage of their study.

During the Implementation stage, educators should be testing the instructions. They can take advantage of a trial project or analogous beta in this stage (Cheung, 2016). Park & Wang (2019) used assessing feedbacks from their trainees, making ready the backup plan, and checking module efficacy, in the implementation stage of their study.

The last stage is the Evaluation stage. In the evaluation stage, educators get feedback about the curriculum and make the fitting changes in accordance with the curriculum (Cheung, 2016). Park & Wang (2019) used analysing feedback about their modules, and preparing quality affirmation tools, in the evaluation stage of their study.

A course taught with traditional methods can be transformed into an online course or a new course may be designed. An example of the implementation of the ADDIE instructional design model to design a new course was for "introduction to software engineering" course planned with a multidisciplinary vision. An adaptation of Programming Languages-I online, which is given on a face-to-face basis at undergraduate level to a fully online course was conducted by Durak and Ataizi (2016).

The most important output of ESCOLA project was the design of the course plan (syllabus) and the content, as the course is given by an online platform. The course plan is designed as an "operating manual" for the platform and the contents.

CREATING COURSE: ADDIE ON PROGRESS

The main target group in this study was engineering HE teachers that include laboratory training and practical applications to their courses. Regarding those teachers, they are open to innovations, highly knowledgeable in their research area, and thus, they might not be resistant to progress and change in the methods of teaching. Thus, the analysis targets them to understand their awareness, readiness, capacity (knowledge and skills), motivation and training needs for adopting ICT-supportive environment into their applied courses. A total of 12 HE teachers from different Engineering

Departments of the Turkish partner took part in the case, providing replies to surveys and interviews explained below.

Analysis Phase

This phase is to understand the stakeholders and their needs. It is aimed at identifying the ICT knowledge and capacity of HE instructors to use in their engineering laboratory courses in order to improve the skills and knowledge of students both in terms of relevant course outputs and in terms of ICT abilities of the students. Thus, the main target group can be defined as HE engineering instructors and students. During the analysis phase, first, the most used ICT tools for laboratory supported engineering education were collected with desk research and feedback from HE lecturers and those tools were analysed. Second, based on these selected tools and some other important aspects of DLMs, the instructors were questioned about their awareness, knowledge, skills, readiness and needs to adopt those technologies to their courses. A mixed method was chosen to collect both qualitative and quantitative data and analysed by experts. At the beginning of the research, a question pool was prepared. After selecting the proper questions by the experts, the researchers established evaluation criteria for them. The ranking was done according to the criteria of using digital learning tools in the laboratory environment. The validity of the measurement tool was assessed according to the language of the tool, suitability of the questions for the purpose and the materials involved. In accordance with the expert opinions, the necessary items were edited by the researchers and the survey was finalised.

The survey was aimed at questioning the instructors for their knowledge level, willingness to use some online tools (digital learning materials) and validity of those tools for their courses. For this aim, 29 common online tools were selected; MATLAB, Autodesk, MIT Open Courseware, Coursera, MOODLE, Photoshop were among the selected online tools. Those tools were categorized as (i) engineering and math design tools, (ii) massive open online courses (MOOC) and (iii) learning materials design tools, respectively. It was observed that 7 of 12 instructors have never heard of, 3 of them have heard of and only 2 of them heard and checked the engineering/math design tools that were listed. Regarding the MOOCs, 8 of them have never heard of, 2 of them heard of but not used and only one of them used in their courses. Similarly, 7 of the instructors have never heard of the digital course material design tools listed, 3 of them heard of but not used, and 1 of them heard and checked and only 1 of them used in their courses. Those results reveal that the instructors from engineering faculty had a lack of awareness and knowledge on the online tools that might be useful for adopting technology in their courses. On the contrary, most of the instructors said to have positive views on using Digital

Learning Materials (DLMs) for their courses and found them useful and helpful for improving learning.

Five participants among those instructors were randomly invited to semi-structured interviews. They were asked to explain if they have used DLMs and how they did it. Videos were the main digital materials that the instructors preferred for their courses, and they did not commonly use additional platforms apart from some engineering software, e.g. AutoCAD, they teach in their courses. Participants from Turkey used particularly videos as supportive digital learning materials for lab courses. Text files, office programs and animations were said to have the other digital materials they used in their courses. Participants from Turkey mentioned economic issues (licence fees) as one of the important limitations of DLMs. Not using often but, most of them had a positive attitude toward using them in their courses (in the future) and believed in its benefits for their students.

According to the instructors, by using digital learning materials for their laboratory courses, they will be able to increase their students' motivation and academic success. By integrating digital technologies more into laboratory courses as supportive materials, the main impact will be in terms of improving the motivation of the students. Those materials will also allow students to practice several times before entering the hands-on laboratory sessions, thus, helping to increase their academic success.

DESIGN PHASE

This phase basically has 4 sub-steps: (a) prerequisites and needed skills, (b) course formats, (c) instructional strategies, (d) assessment approaches. Design phase "coincides with the development of a plan for the learners to achieve the objectives".

Prerequisites and Needed Skills

The objective of the training course is to increase the abilities of Engineering Higher Education Instructors to pursue innovative teaching strategies supported by the intelligent use of digital learning platforms and tools.

By the end of this training course, the participant will be able to:

1. Recognize the key terms in digital education,
2. Tell the importance of the use of digital tools and learning platforms in education,
3. Develop strategies to innovate their teaching strategies,
4. Explain/clarify the importance of digital inclusion,

5. Identify the technologies, tools and methods useful for specific classes,
6. Define the benefits of online learning,
7. Explain the benefits of video and simulation techniques for practical laboratory applications,
8. Understand the importance of using virtual learning platforms in the current education system,
9. Implement digital tools and learning platforms for/in their courses.

According to the objectives mentioned above, there should be decided crucial prerequisites and needed skills. During the analysis process, prerequisites and needed skills are defined as:

1. Being an engineering instructor
2. Teaching engineering courses
3. Having strong engineering terminology
4. Knowing the basic jargon of digital technology
5. Being aware of possibilities to use digital technologies in classroom settings
6. Willing to share the knowledge
7. Being ready to learn

Those basic prerequisites and needed skills defined for instructors are to teach their colleagues to understand basic needs before starting to put other bases of the design phase. Since both sites of teaching process are instructors, it is better to design the whole course under those skills.

Course Format

This course was decided to build in an online learning environment to reach more participants. Therefore, all materials and learning techniques were considered by the online learning environment. "https://escolaproject.eu/" URL address was taken to put all digital learning materials into that learning platform.

Instructional Strategies

Instructional strategies determined after deciding course format. Since it was an online learning platform, it should be aimed to select adaptable online instructional strategies.

Assessment Approaches

It was a critically important decision to determine what kinds of assessment approaches should be chosen. There were basically two different approaches to employ: (i) summative and (ii) formative. It was decided to use both assessment approaches to measure the learning process effectively.

Development Phase

This phase is all about setting the platform for the first pilot run. The 'ESCOLA – Digital Teaching Tools for Engineering Labs' training course is a flexi-time course and is delivered through e-learning as a training delivery method. The course duration is approximately 30 hours. At the design phase, the course consisted of 5 modules. With the feedback taken during the implementation of the course, one more module was added. The modules started with a short introduction on the topic of the module including the objective and expected learning outcomes. In the main part, the module content was supported by practical examples. Every module is concluded with a summary and provides a self-assessment test. Content is given on the online platform "https://escolaproject.eu/" with below (See. Table 1);

Implementation Phase

A learning activity was designed for the implementation phase and the training modules were explained to them one by one with supporting exercises given at one of the partner's premises in Szczecin, Poland. Each education partner selected two participants for the learning activity among the HE instructors that teach laboratory supported courses. The main criteria for selection were readiness, willingness and basic knowledge on using ICT. After the learning activity, they were asked to implement the knowledge they gained from ESCOLA in at least one of their laboratory classes and they performed a pilot test with their students. Two instructors were selected from Turkey for the implementation of ESCOLA training course, and two others were selected from the other two universities of Bulgaria and Poland. In this chapter we focused on the knowledge and skills that the instructors from Turkey gained by this experience.

Such training activities are supported by university administrations to support life-long learning and to enhance the teaching performance of the instructors. It is of great importance to raise the awareness of instructors on the development of instructional technologies. Compared to instructors, students are more likely to adopt those technologies and the engineering students are one step forward. Thus, the course was given (face-to-face) to selected two engineering instructors in the

Table 1. Course modules

Module Title	Objective	Units (subtitles)	Expected Learning Outcomes
Digital inclusion	• to get an introduction into new technologies and to use them for promoting digital equality in education	• Aspects of digital inclusion • The route to digital inclusion • Key goals for digital inclusion • Digital Intelligence • Summary	• Understand that digital inclusion is more than just access to the internet and the skills to use it • Explain the importance of digital inclusion • Clarify the concept of Digital Intelligence
Innovative teaching	• to get an introduction into innovative teaching methods and the new technologies that are developing in the world of learning in engineering and how teachers can use those for promoting educational equality.	• Trends in education • Innovative teaching methods • Characteristics of being an innovative educator • Innovative use of instructional technologies • Technology-enabled laboratory-based teaching and learning • Summary	• Get to know new teaching technologies and methods • Be inspired by emerging trends in education • Learn how to be in the mindset of an innovative educator • Gain insight into techniques and tools for educational innovation • Identify the technologies and methods useful for specific classes
Distance learning	• to introduce the term of distance learning, meaning how you can learn on the move, with or without handheld technology	• What is distance learning? • Usage scenarios in distance learning • Forms of distance learning • Examples for distance learning • Summary	• Identify the benefits of distance learning • Interpret/tell the modern use of educational technology facilitates online learning and independent learning by the widespread ICT usage • • Select the right ICT tools that can be used in everyday classroom practices for transforming teaching and learning
Videos and 3d simulations for laboratory	• To integrate video and 3d simulations as learning materials for engineering laboratory teaching.	• Digital teaching tools for engineering laboratories • Video and simulations for engineering teaching • Videos • Remote Labs • Simulations (Virtual Labs) • Summary	• Awareness on simulations and videos for engineering laboratory teaching • Understanding the benefits of videos and simulations for laboratory practical applications • Ability to select suitable video and simulation techniques for relevant laboratory course
Collaborative learning in virtual environments	• to help engineering teachers adapt to the new collaborative virtual learning spaces.	• Collaborative Learning Spaces • Key features of Virtual Learning Environments • Types of Virtual Learning Environments • Collaborative Learning in Virtual Environments • Creating a Collaborative Virtual Learning Environment • Summary	• Discover different virtual learning platforms and their key features • Understand the importance of using virtual learning platforms in todays' education system • Learn how to create an impactful virtual learning environment

training activity in Poland, two more instructors were selected for further training, they took the modules online, and completed the quizzes. Those four teachers from Turkey were asked to implement the ESCOLA approach in one of the laboratory courses they gave.

Evaluation Phase

The instructors that were given the ESCOLA training applied a sample of the ESCOLA training course with their students and took responses from their students as well. Based upon their own experience and the feedback from students, the participating teachers provided their comments and approaches via a survey with open-ended questions:

1. The part of the learning programme and platform they found the most useful and interesting and state their reasons for their response.
2. If they think that addition or revision is needed in the programme content, activities and resources and explain their reasons for their response.
3. If they recommend ESCOLA learning to other teachers along with their reasons.
4. If they think the participation in the training programme has changed their skills and competences and attitudes towards their teaching approach.
5. If they think that they will apply the issues they learned in this course regularly in their teaching/training activities.
6. If they feel a change in their teaching approach that impacts their students in terms of motivation, skill acquisition and labour market preparation.

Following cases were analysed in the evaluation phase in order to use for the revised design of the course;

Case I- Instructor in Food Engineering, Teaching Experience: 21 yrs.,

"I learned the ESCOLA innovative approach in the training course in Poland. I learned how to integrate instruction technologies to my laboratory courses, I did not have much difficulty in applying the innovative approach in my test course, but I will try to diversify my all course materials with the new knowledge and approaches I learned.

I have used the online tool "mentimetre" to question their approach to concepts such as "biotechnology" and "fermentation" and related "laboratory tests" we discussed on those terminologies. My aim was to improve the interactivity of the course by taking the students responses.

While using this approach, I experienced the use of many internet resources for my courses, I experienced a change in my approach giving my courses. This change caused the students to participate more in the course. I think this approach is definitely student-centred because the students themselves put more effort to contribute to the course, stay more focused during the course and learn better."

In the short conversation, I made with my students at the end of the lesson, they stated that they were very pleased with the technology integrated lesson and they learned easily without getting bored."

Case II- Instructor in Bioengineering, Teaching Experience: 16 yrs.,

"In a group of 10 students, I used the brainstorming technique, I expected the students to answer one question as soon as possible with a short sentence by giving the next round to the next student. With this technique, the written ideas give birth to new ideas with an increasing impact.

I used this technique while teaching reactor design and kinetics.

I liked this innovative approach because it was practical and applicable, and I could teach more effectively. Thanks to the training I received under ESCOLA, I knew in advance that I wouldn't have any difficulties. I realised that the students were motivated to the course much more easily. At the same time, they took advantage of each other's ideas and new ideas and very good approaches developed at the end. Before applying this approach, I informed my students about ESCOLA. I have expressed my intention to apply such innovative approaches in all my courses. My students gave positive feedback about the use of such practices in the other courses. Since I believe that this approach gives positive results, I decided to adopt what I think is appropriate from the other techniques I learned within ESCOLA to my teaching experience."

Case III-Instructor in Civil Engineering, Teaching Experience: 11 yrs.

"I used the Kahoot tool to test my students' knowledge at the laboratory, it was easy to perform and nice to test their knowledge during their lab experience. The course entitled "Soil Improvement Methods", elective course for final grade undergraduate students. The students were motivated, and they thought it was fun. By adopting this approach to my courses, I found out that I can evaluate my students' knowledge by techniques other than classical open-ended questions. I feel motivated for learning other tools for adopting instructional technologies to my courses."

Case IV-Instructor in Civil Engineering, Teaching Experience: 12 yrs

"I have used EDMODO as a teaching management system, we have cast some videos of laboratory applications and edited them by using CAMTASIA with my students as a part of a project. I used EDPUZZLE application to share the videos with my students. Students came to hands-on laboratory application after watching the videos. I have asked them some questions regarding those tests in the Quizzes and they have prepared laboratory reports.

The main challenge for me was preparing new course materials and the adoption of course content according to the new techniques. It took plenty of time to prepare new materials and make changes in the teaching methodology, but since the system is ready now, it will be easier next year and so on.

It was interesting for me to learn that there are many technological tools I can use to support my courses. Now I can see that when you have learned how to use those tools, it becomes easier to implement them to other courses. I realised that my students are better at telling what they have learned in the laboratory, they remember more on the lab tests and they have better grades. At the beginning they were not as motivated as they thought, watching the videos was an extra duty for them. When preparing the lab reports and working for the exams, they realized that the videos were a good source for them to remember what they have done in the lab."

Analysed views of participants show that the instructors have used the ESCOLA approach with the different tools and techniques they have learned. As they were free to select their way of implementing this approach, two of them choose to apply for their laboratory courses, one decided to use it for design approach (brainstorming) and the other used this to reach and interact with their students while putting them into the centre of the learning process. All those experiences had a common conclusion; the students were happy (felt better, motivated, more engaged) at the end and the instructors would like to integrate this student-centred innovative approach more to their teaching experience in the future.

TERR MODEL

We would like to introduce our Experiencing on Professional Development Process as a step-based model which is called TERR.

TERR was developed by authors to demonstrate the professional development (PD) process. We define TERR model as PD model which gives basic instruction of the path that followed during the training. TERR model has own steps which come with acronym of those steps such as Train, Evaluate, Reflect and Revise (See. Figure1). Each step supports trainees' connection to the training, shows how this task is related to training by evaluating, makes them feel how important the task is by reflecting their thoughts, and makes them sure how their learning and the learning process is getting more effective by revising the training with their reflection.

Figure1. TERR model

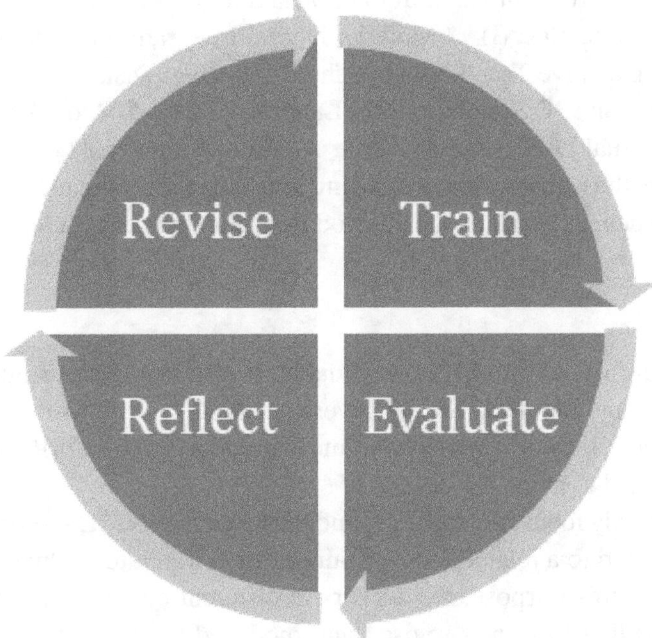

TERR model shows the real-time steps of ESCOLAOnAir Training. Those steps were planned by the research team and practised in EscolaOnAir training. Those steps, (i) train, (ii) evaluate (iii) reflect and (iv) revise.

Train

Train (T) defines the learning action itself. The course design based on ADDIE should be completed, the pilot run should be finished, and all evaluations process should be done before starting T in TERR. T is to deliver real-life actual course. For that purpose, EscolaOnAir training was organized. EscolaOnAir is a one-day professional development training. 16 academicians participated the training. The participants were informed about the event through social media, e-mail and on websites such as the conference entitled "International Instructional Technologies in

Engineering Education". The participants were selected from various specialisations including different branches of engineering (bio, chemical, textile, civil).

Since participants study on different fields, the courses they teach vary and it would be a long list with a wide range of course variety. They teach courses at any level in universities from undergraduate (freshman, sophomore, junior, senior) to graduate levels.

The training started with an introduction and followed from the 1st module to 5th module. Those modules: (i) 1st Module: Digital Integration in Education, (ii) The 2nd Module: Innovative Teaching, (iii) The 3rd Module: Distance Learning, (iv) 4th Module: Videos and 3D Simulations for Laboratories, (v) 5th Module: Collaborative Learning in Virtual Environments. Some of the innovative tools were demonstrated to the audience thus, an awareness on using innovative teaching tools was raised and this helped audience to understand the modules better thanks to the enriched content.

Evaluate

Evaluate (E) defines as proving how valuable this training is by evaluating training and their learning outcomes. Step E covers small quizzes to understand what they learned and how they learned. Those measurements give information about the training itself and content.

This phase only focuses on training and the process itself. Therefore, assessment tools were created to a related focus. Trainees were asked to evaluate each module of training. For this purpose, researchers in different countries created a form of evaluation and that form was revised and modified by each country considering cultural reflections. The form shows that general evaluation of the training is successful. Therefore, trainers and instructional technologies should focus on where trainees find "Very Poor" and "Poor" side of the training.

The form shows that some modules found that they needed to be fixed and modified. The next step of TERR is also to concentrate on how those items were found to be fixed by trainees via gathering their reflections to understand the issues or the reasons for those challenges in depth.

This step basically shows the pros and cons of the whole training itself which should be aimed to improve the strengths and revise the weaknesses according to trainees' views on training. Therefore, it is needed to have trainees' reflections.

Reflect

Reflect (R) defines as being a part of the training process. This step is to give a privilege to the trainees to change the training process. Increasing belongingness for learning process is critical to raising the importance of not only learning process but also the training process itself. The interviews with trainees, open-ended questionnaires, and small discussions might help to get their reflections on both learning and training process.

Reflecting phase is to put trainees' views together for understanding what as a trainer or an instructional technologist should be revised. As highlighted in the evaluation phase, couple of contents mentioned to be fixed during the training. It is important to get into detail and see the what trainees think about them. On the other hand, this process also shows how trainees reflect in a positive way.

One of the trainees said, "I think module 4 is the most effective, the content is related to the branch of my department, I found the laboratory practice and simulations very important for my courses." One of other trainees also mentioned "I found Module 2, the innovative teaching part the most effective. Because different teaching methods and application techniques were mentioned". Those reflections showed how the training worked. However, some of trainees highlighted that Module 1 Digital Innovation didn't meet their expectation and they didn't find it interesting. It was also talked about that Module 4 wasn't that interesting as expected. This process shows not only pros and cons but also what it is needed to improve your training, something that you haven't thought before such as adding new modules, simplifying the jargon of context, adding more example to visualize data and so on. For this purpose after the feedback taken from the trainees, it was decided to include one more Module that introduces some of the most used online (and free) tools that they can use to prepare digital content for their courses, and some supplementary papers and e-books to enrich their knowledge on these technologies.

Revise

Revise (R) defines as making training better by reflecting trainees' views. Revise might affect to make changes not only in T step in TERR but also in Step1 and Step2. Revision is to the last step of TERR to make necessary changes to improve training.

At the last step of TERR, it becomes revised by making changes as per the trainees' expectations. This process does not only make your process better but also put the trainees' thoughts for changing the training and make their voice heard. That increases their belongingness towards training. At that point, Module 1 and Module 2 was modified and filled the gap between training and trainees' expectations. Moreover, a new module was added as a 6th module as requested. These new modules included hands-on and applicable tools and their examples.

TERR model is a model that might be used in every training to understand how your process works. It should be considered that some steps could be jumped. For instance, nothing needs to be changed and after the reflecting steps, the revising step might be jumped over. It also should be noted that TERR model worked during our training. We are still working on this model to increase its validity by applying it to different trainings.

In conclusion, TERR model gives basic road maps to trainers for their training process. It should be supported and practised by experts in the field of education.

CONCLUSION

This chapter explains two main topics with our experiences. First topic is about creating a course by using ADDIE. Second topic is about employing TERR professional development model. We call our sharing process Create, Implement and Reflect. During the application of ADDIE model, we create a course by following its own steps. Therefore, this was a creation phase of our sharing experience. After the completing the course, we applied in a real-time course (ESCOLAonAir). This process showed us how implementation might work with hands-on experiences. This implementation phase is the starting point of TERR model. TERR model was model which was shaped with our real-time experiences to demonstrate our professional development process. TERR model was the last stage of showing our reflection for whole process. TERR model was employed only our professional development training. Therefore, TERR need to be used in different professional development trainings to validate itself.

Those steps above were followed for each partner during the ERASMUS+ funded project. However, those steps might be changed, modified and re-considered by partner countries due to cultural differences. For that reason, this chapter covers only Turkish side of process. We shared our experiences during the training process. It might give some applicable and feasible path for researchers to enhance and facilitate their own processes.

ACKNOWLEDGMENT

The authors would like to acknowledge the Engineering Student Centered Learning Approaches (ESCOLA) project funded by the European Union and coordinated by Turkish National Agency under Erasmus+ Programme. (Grant number 2017-1-TR01-KA203-045955).

REFERENCES

Branch, R. M. (2009). Instructional design: The ADDIE approach. Springer Science & Business Media.

Cheung, L. (2016). Using the ADDIE model of instructional design to teach chest radiograph interpretation. *Journal of Biomedical Education, 2016*, 1–6. doi:10.1155/2016/9502572

Durak, G., & Ataizi, M. (2016). The ABC's of Online Course Design According to Addie Model. *Universal Journal of Educational Research*, *4*(9), 2084–2091. doi:10.13189/ujer.2016.040920

Englund, C., Olofsson, A. D., & Price, L. (2017). Teaching with technology in higher education: Understanding conceptual change and development in practice. *Higher Education Research & Development*, *36*(1), 73–87. doi:10.1080/07294360.2016.1171300

Ossiannilsson, E. (2018). Promoting active and meaningful learning for digital learners. In Handbook of Research on Mobile Technology, Constructivism, and Meaningful Learning (pp. 294-315). IGI Global. doi:10.4018/978-1-5225-3949-0.ch016

Park, M., & Wang, L. (2019). Instructional design of technology-enhanced process writing for secondary EFL learners in Hong Kong. In J.-B. Son (Ed.), Context-Specific Computer-Assisted Language Learning (pp. 122-136). Queensland, Toowoomba, Australia: Asia-Pacific Association for Computer-Assisted Language Learning (APACALL).

Peterson, C. (2003). Bringing ADDIE to life: Instructional Design at Its Best. *Journal of Educational Multimedia and Hypermedia*, *12*(3), 227–241.

Chapter 4
Educational Data Mining:
A Systematic Literature Mapping Study

Aslıhan Tüfekci
Gazi University, Turkey

Esra Ayça Güzeldereli Yılmaz
https://orcid.org/0000-0002-2574-1628
Afyon Kocatepe University, Turkey

ABSTRACT

The education-training process and all activities related to it have the power to direct the future of societies. From this point of view, the process should be analyzed frequently in terms of input, output, and other process elements. Educational data mining is a multidisciplinary research area that develops methods and techniques for discovering data derived from various information systems used in education. It contributes to the understanding of the learning styles of learners and enables data-driven decision making to develop existing learning practices and learning materials. The number of academic and technical research on educational data mining is on the rise, and this has led to the need to systematically categorize the existing practices. This systematic mapping study was conducted to provide an overview of the current work on educational data mining and its results are based on 153 primary sources including journal papers, articles published in magazines, conference and symposium papers, theses, and others.

DOI: 10.4018/978-1-7998-2562-3.ch004

INTRODUCTION

Technological developments that are now influential in educational environments have also enabled large volumes of data to be generated in the field of education with the integration of technology into education. The contribution of this data set to the quality of education is directly related to the uncovering of meaningful patterns within the data. For a quality education, higher education institutions should be able to make the right decisions in administrative and educational sense (Tufekci & Guzeldereli, 2018). Problems such as inadequate academic planning (course determination, content determination, course credit assignment, etc.), failure to predict students who may fail, and students who have the potential to leave school are among the problems of higher education institutions. Producing solutions for these problems can be achieved by in-depth analysis of existing data (Şengür & Tekin, 2013). Thus, some measures can be taken to increase the quality of education.

The educational process and all activities related to this process have the power to direct the future of the society. From this point of view, it can be said that the process should be analyzed frequently in terms of input, output and other process elements. Although this analysis is carried out at micro and macro level achievement exams, the convergence of the achieved success to the desired one is controversial when the exam scores are considered as the only input. Therefore, it is important to predict the transition period of input to the desired output in order to establish the awareness of the situations in which interruptions in the process should be intervened. In this respect, educational data mining methods can be a powerful tool for academic interventions (Tufekci & Guzeldereli, 2018).

Educational data mining is concerned with issues such as developing recommendations for students, providing feedback and support to trainers, modeling student data, grouping students, supporting planning activities, and identifying student analysis. Educational data mining is a relatively new field of study in educational research and it increases its importance foreducators day by day. It is an interdisciplinary field of study which is directly related to many areas such as computer science, statistics, mathematics, data visualization, etc. It aims to transform the data produced by the information and communication technologies used in education into meaningful information for actors involved in education by analyzing them with various methods (Güldal&Çakıcı, 2017). When the studies conducted in the field of education are examined, data mining seems to be used for the purposes of classification, clustering, cohesion rules, methods and techniques for students' achievements, clustering according to their information, determination of their interests and trends, automatic presentation of learning contents and revealing misconceptions (Tufekci & Guzeldereli, 2018).

In this study, it is aimed to get an overview of the systematic classification and mapping of current studies on educational data mining.The contribution of the study is the systematic mapping of primary studies in the online resource repository for educational data mining.Thus, it is possible to identify gaps in the field and to develop new approaches in this direction.The primary objective of this study is to reveal which techniques are mainly used in the studies, which data mining techniques are applied in educational environments, and what kinds of contributions these studies provide to the literature.In the following sections, information is given about educational data mining and its applications and general systematic mapping process, followed objectives and research questions, research method are explained in detail.The selection criteria of the sources included in the study and the process of formation of the mapping pool were discussed and the iterative development of the map was analyzed.Finally, the results obtained from the systematic mapping and the studies that are planned to be done in the future are included.

BACKGROUND

With the use of today's technological systems in education, it has provided large data sources in the field of education. Data mining practices are based on the fact that there is a lot of undisclosed confidential information that can be useful both for students, teachers, administrative staff and educational institutions in large data stacks of the educational process. With the development of technology, it is foreseen that future educational environments will be designed with hidden patterns to be obtained from these data.

There has been an increase in the number of data mining studies particularly in the areas of health, marketing, banking, finance, stock exchange and the internet. Data mining studies in the field of education are still under development. In particular, IJAIED, International Conference on Intelligent Tutoring System (ITS), International Conference on Educational Data Mining (EDM), Journal of Educational Data Mining (JEDM), International Conference on Data Mining for User Modeling (DMUM) contributed to this field with many conferences and articles (Barahate, 2012). The studies in this area are based on academic achievement, factors affecting success, student profile, how to help students failing and estimation of GPAs Educational data mining is a relatively new field of study in educational research and it increases its importance among educators day by day. It aims to transform the data produced by the information and communication technologies used in education into meaningful information for actors involved in education by analyzing them with various methods(Özbay, 2015).

Today, a lot of information related to students' personal information, their grades, the subjects in which they are successful and unsuccessful are kept in large databases. These masses of data, from which significant relationships can be searched and where important information can be obtained, can be used in determining the problems that cause disruption in education, in predicting student achievement, adapting learning environments to student characteristics, identifying students with a tendency to leave the course and increasing the quality of education in this direction (Kotsiantis, 2009).

In this study, it is aimed to systematically map educational data mining related studies. Thus, it will be possible to draw conclusions such as the general tendency of the studies produced in this field, the techniques used to a large extent in the field, and what kinds of contributions the studies provide to the literature.

RESEARCH METHOD

In this section, an overview of the formation criteria of the systematic map and mapping questions are discussed.

The Purpose of the Study and the MappingQuestions

The aim of this study is to make determinations about the current studies in the field of educational data mining, to discover which orientations they are in and to determine the existing gaps for future research. For this purpose, the literature pool obtained from the studies developed on educational data mining applications has been systematically mapped and revised in order to find current approaches and trends in the field. The following mapping questions (MQ) have been created for this purpose:

- MQ 1- Examination of the studies examined within the framework of systematic map according to the type of contribution it provides to the field: Method, technique, model, tool, process etc. of the scientific publications examined in the field of educational data mining. In terms of the answer to the question of what kind of contribution it makes to the literature. Contributing to mapping by addressing this question will help identify gaps in current work and understand the overall trend of the area.
- MQ 2- Examining the studies examined within the framework of the systematic map according to the research method: With this question, an answer is sought to the question with which research method the studies in the literature pool are developed. The studies examined to respond to MQ 2

were categorized according to six different research methods and each study was classified to be included in only one method. These types of research methods are; basic research, applied research, experimental development, product development, descriptive research and others.

- MQ 3- Classification according to the educational data mining techniques used in the studies examined within the framework of the systematic map: The studies included in the mapping process are examined and each of them takes its place in the map with the educational data mining technique it uses. Thus, the density of the techniques used on the map can be seen, and the least preferred techniques can be obtained.
- MQ 4- In the studies analyzed within the framework of systematic mapping, classification according to the purpose of using educational data mining techniques: What and how the educational data mining techniques are used is tried to be answered at this stage. Getting the answer to this question will help to understand what the overall trend of the studies is.

An Overview of Systematic Mapping Process

The formation of the systematic map developed in this study is summarized in Figure 1 with three main stages. These stages were determined based on the studies of Peterson et al. (2008).

- Selecting a publication for the literature pool
- Developing a systematic map based on criteria
- Results and evaluation from systematic mapping

Selecting a Publication for the Literature Pool

The first stage of the systematic mapping process is to determine the resources that will be on the map. In this process, the following steps are followed:

- Selection of publications with keywords
- Setting inclusion and exclusion criteria for mapping
- Termination of the literature pool

Figure 1. Protocol scheme used in systematic mapping

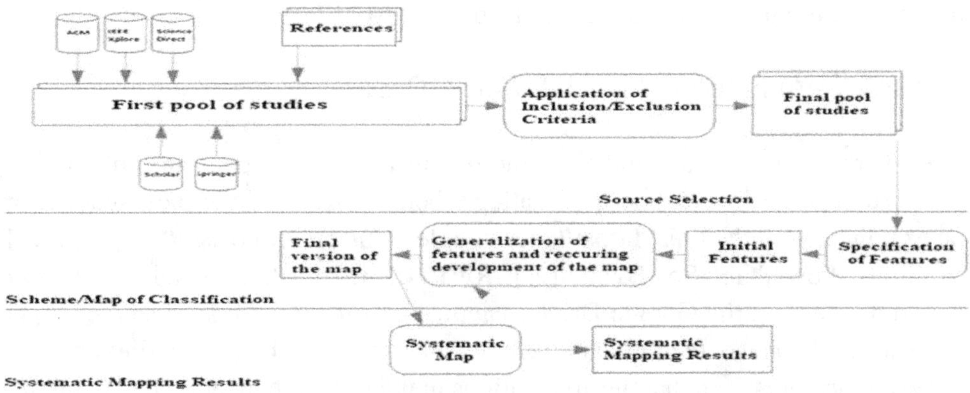

Phase1. Selection of Publications with Keywords

IEEE Xplore, ACM Digital Library, Science Direct, Springer and Google Scholar libraries have been determined to find the publications that will be included in the systematic map. Keyword combinations such as "educational data mining", "data mining in education" and "education and data mining" were applied on these libraries, and studies in the literature pool were obtained. Theses, articles and book chapters were taken into consideration in obtaining these studies. Conference papers were not included in the mapping study. In the literature pool obtained in the first stage, a total of 168 studies were included, and after the review of the summary and introduction sections of the studies, their relations with the field were evaluated and the number of studies in the literature pool was reduced to 153.

Phase 2. Setting Inclusion and Exclusion Criteria for Mapping

In this study, a number of limitations were applied to the creation of the resource pool. The first of these is the direct relevance of each study to the field of educational data mining, the level of consistency, evaluation and validity of the studies. The second is the inclusion of studies written in English or Turkish and accessible electronically. In order to apply the inclusion and exclusion criteria at the initial acquisition of the literature pool, each study was subjected to a rating. In this context, the studies are graded as "1" and "0". "1" here indicates that the study can be included in the literature pool, and "0" indicates that the study will be excluded. The evaluation of each study was done by first reviewing the title of the studies, then the summary and keywords. If these are not sufficient, a more detailed assessment is made and the introduction and sub-sections are reviewed. Ultimately, the number of studies in the literature pool has been reduced from 168 to 153.

Phase3. Termination of the Literature Pool

The reference list of 153 studies in the obtained literature pool and the grading information according to the classification scheme described in the next section can be examined with the link "https://docs.google.com/spreadsheets/d/1E2pBX5-2ahWYKWlUFow58GbtzSoAB2hAhLUjflRwT54/edit#gid=0". This list has been stored online with the Google Drive system. The annual volume of studies in the field of educational data mining is presented in Figure 2. When the studies obtained are examined, it is seen that the applications in the field of educational data mining started to appear in the beginning of 2000, and it has shown an increasing density since 2013.

Figure 2. Year-based distribution of studies

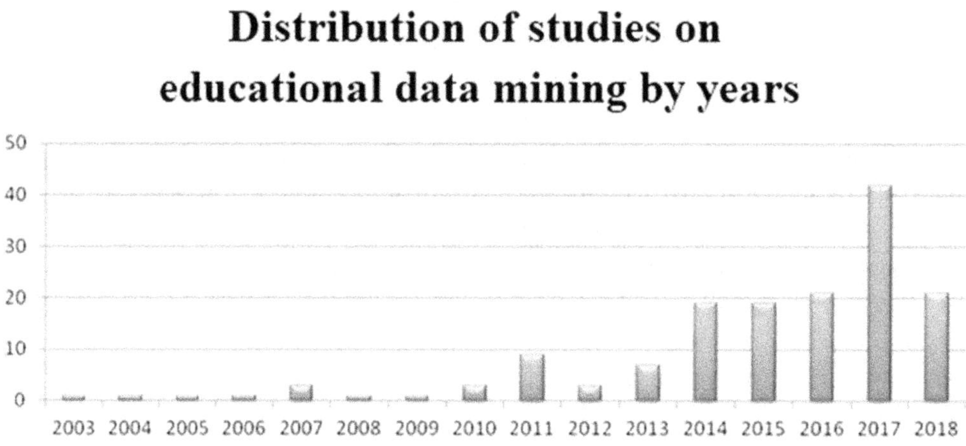

Developing a Systematic Map Based on Criteria

After completing all these steps, the classification scheme to be used for the classification process is given in Table 1. The first column here is the mapping question (MQ) list. The second column is related to the subject to be classified. The third column shows all the possible values for each feature. Finally, the fourth column relates to whether a feature can be selected more than once for each run. In determining these criteria, Tufekci et al. (2018) has been benefited from.

Table 1. Scoring criteria for the risk-related impact factor (Tufekci et al., 2018)

MQ	Attributes	Categories	(M)ultiple/ (S)ingle
1	Type of contribution	{Methods / Techniques, Models, Tools, Processes, Case studies, Others}	M
2	Type of research	{Basic research, Applied research, Experimental Development, Product development, Descriptive research, Others}	S
3	Type of data mining technique	{Association analysis, Clustering, Classification, Estimation by ANN, Others}	M
4	Purpose of educational data mining	{Identification, Prediction, Knowledge discovery, Comparison, Others}	S

Results and Evaluation from Systematic Mapping

In order to obtain a systematic map, the results obtained with the questions directed to the studies in the pool are presented in this section.

MQ1- Examination of the Studies Examined on the Systematic Map According to the Type of Contribution to the Research Area

As mentioned in the previous sections, the first of the mapping questions is concerned with the meaning of the studies contributed to the literature and is to group studies as educational data mining techniques, models, tools, processes, case studies and others. Figure 3 shows the contribution type distribution of 153 sources in the literature pool to the research area.

It is seen that most of the studies examined in Figure 3 contribute to the educational data mining process. In addition, it was determined that 73 studies evaluated contributed to the research area with new techniques and models. In addition, it is seen that there are 38 studies developed with the application of techniques on the case and 19 studies contributing with the new tool developed in the field.

Figure 3. Classification of the studies examined according to the type of contribution to the research area

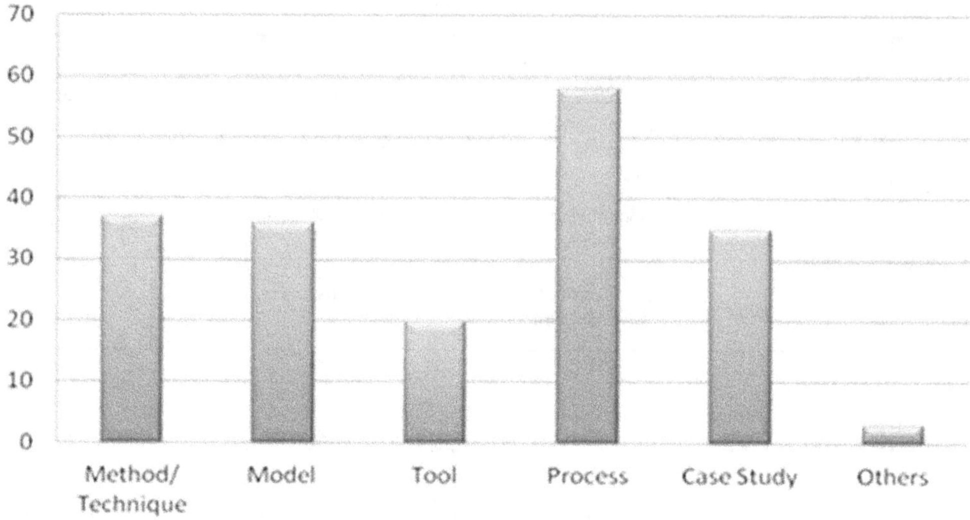

MQ2- Examination of the Studies Examined on the Systematic Map According to the Research Method

In order to determine what kind of research methods are used in the studies examined, the second thrust question was directed to all studies. Accordingly, Figure 4 shows the distribution of the studies in the literature pool in terms of research type. Upon asking this question to the studies examined, it can be seen in Figure 4; The most commonly used research method is basic research. There, we can think that time and cost factors are effective. In addition, 52 studies were found to be based on a research question or developed based on a hypothesis. It is seen that many studies have been developed to apply data mining techniques on educational data in the field of education. However, the number of studies focused on product development is very low. In addition, it is one of the conclusions to be drawn here that explanatory researches in which the suggestions for the future are developed have been frequently seen in recent years.

Figure 4. Classification of the studies examined according to the research method

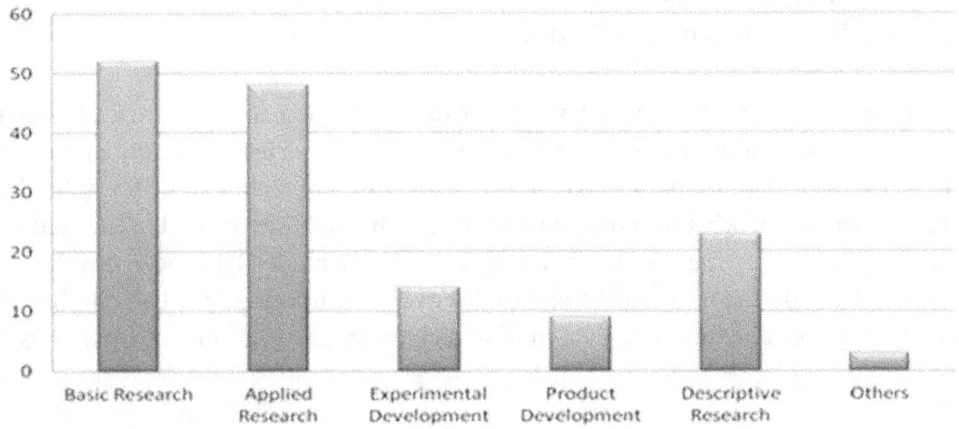

MQ3- Classification According to the Educational Data Mining Techniques Used in the Studies Examined in the Systematic Map

Along with the third mapping question of the study, the studies in the literature pool are classified according to data mining techniques to examine which data mining techniques are used extensively. Accordingly, Figure 5 shows the distribution of data mining techniques used in 153 studies examined. In the evaluation made in this context, it is seen that clustering and classification techniques are used intensively, and studies using the association rule analysis technique are relatively less.

Figure 5. Classification according to the educational data mining techniques used

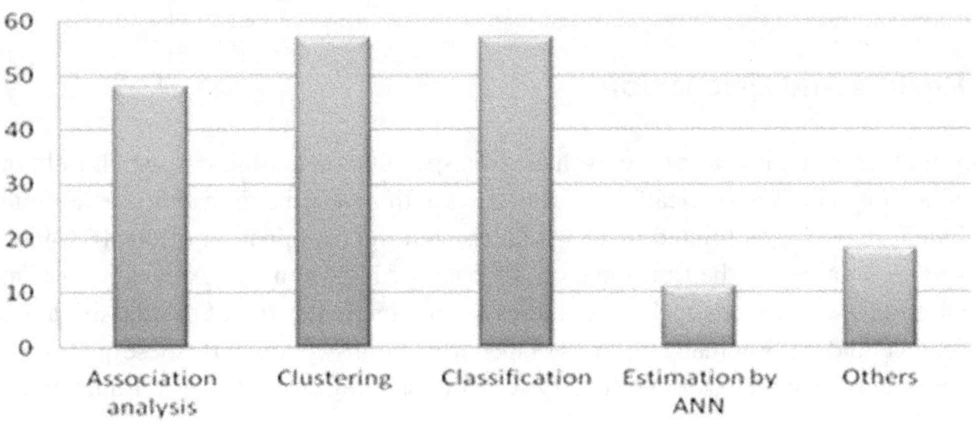

MQ4- Classification According to the Purpose of Using Educational Data Mining Techniques in the Studies Examined in the Systematic Map

The last mapping question aims to reveal the data examined on the purpose of which data mining techniques are used. In this context, all studies in the literature pool are classified according to the purpose of using data mining techniques. Accordingly, Figure 6 shows for what purposes data mining techniques serve in all of the studies studied. If the assessment is made; It can be said that studies have been developed to predict and define information and to discover new information. The number of studies developed to compare data mining techniques is lower than other studies.

Figure 6. Classification according to the intended use of data mining techniques in education

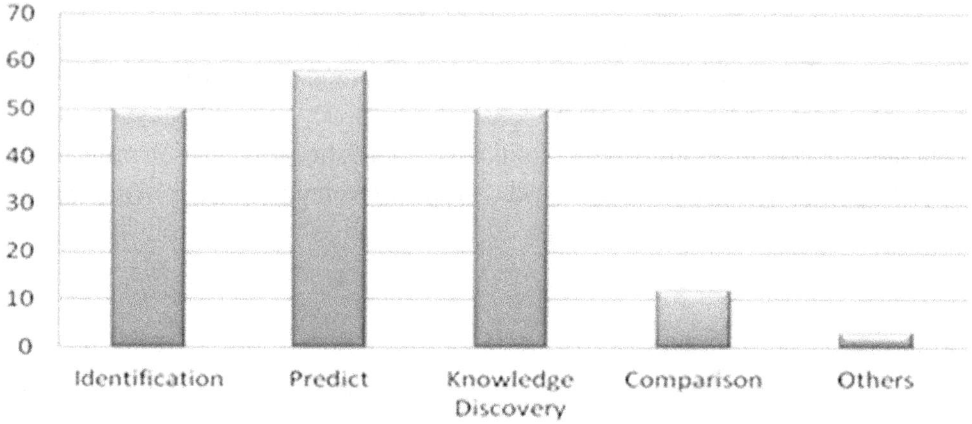

Results And Discussion

With this study, it is aimed to evaluate all aspects of data mining research field in education and to numerically characterize it with a systematic mapping. The final literature pool was limited to 153 studies, with the examination of the 168 main studies obtained in the first stage on the basis of inclusion and exclusion criteria. Questions were asked to all these studies within the framework of the classification scheme, and the systematic map was shaped with the answers given to these questions. The results obtained as a result of systematic mapping are as follows; Educational data mining studies started to appear in the 2000s, and it has been observed that their numbers have increased since 2013. It has been determined that most of the studies examined contributed to the application of data mining techniques. On the other hand, it is clear that there are very few studies on product development that reveal new tools. When the studies were evaluated according to the data mining techniques used, it was observed that the classification and clustering techniques were mostly studied, but the publications working on future predictions were rarely compared to these studies. With the expansion of this study, it is aimed to further increase the accuracy levels of the results obtained and to assist in determining the gaps in the field for future studies.

When the map obtained from the study and the graphs obtained from the map are examined, it can be predicted in which ways the studies in the field of educational data mining are incomplete.For example, in most of the studies, it is seen that the tools that provide solutions by operating data mining algorithms have not been developed, but the studies are supported through ready-made commercial software.However, special tools designed based on the relevant data structure and field will contribute to the field by producing more efficient results.In addition, the fact that the studies have the structure to reveal hidden relationships and produce determinations, and the fact that the number of studies that include forecasting for the future is quite small is considered as the missing part of the field.The transformation of existing data into predictions that will shed light on the future can be transformed into a number of applications which, together with the field of artificial intelligence, give very efficient and effective results.The fact that the studies to be developed in this context is based on systems that make predictions and can work together with intelligent decision support systems can contribute greatly to the field.

REFERENCES

Baker, R., & Associates. (2010). Data mining for education. International Encyclopedia of Education, 7(3), 112–118.

Barahate, S. R. (2012). *Educational data mining as a trend of data mining in educational system.* International Conference & Workshop on Recent Trends in Technology (TCET), Mumbai, India.

Buniyamin, N., Mat, U. B., & Arshad, P. M. (2016). Educational data mining for prediction and classification of engineering students achievement. In *IEEE 7th International Conference on Engineering Education*. IEEE.

Calders, T., & Pechenizkiy, M. (2012). Introduction to the Special Section on Educational Data Mining. *SIGKDD Explorations, 13*(2), 3–6. doi:10.1145/2207243.2207245

Güldal, H., & Çakıcı, Y. (2017). Educational Data Mining, Balkan. *Educational Studies, 1*, 135–143.

Kotsiantis, S. (2009). Educational Data Mining: A Case Studyfor Predicting Dropout-Prone Students. *International Journal of Knowledge Engineering and Soft Data Paradigms, 1*(2), 101–111. doi:10.1504/IJKESDP.2009.022718

Özbay, Ö. (2015). Data Mining Concept and Data Mining Applications in Education. *International Journal of Educational Sciences, 2*(5), 262–272.

Özbay, Ö. (2015). The concept of data mining and data mining applications in education. *International Journal of Educational Sciences*, (5): 262–272.

Özdemir, Ş. (2016). *An application on data mining in education and predicting student academic achievement* (PhD Thesis). Istanbul University, Institute of Science and Technology, İstanbul, Turkey.

Petersen, K., Feldt, R., Mujtaba, S., & Mattsson, M. (2008). Systematic Mapping Studies in Software Engineering. *EASE, 8*, 68–77.

Romero, C., & Ventura, S. (2007). "Educational Data Mining: A Survey from 1995 to2005". Expert Systems with Applications,Cilt: 33. *Sayı, 1*, 135–146.

Şengür, D., & Tekin, A. (2013). Estimation of Graduation Grades of Students with Data Mining Methods. *Journal of Information Technologies, 6*(3).

Thomas, E. H., & Galambos, N. (2004). What satisfies students? Mining studentopinion data with regression and decision tree analysis. *Research in Higher Education, 45*(3), 251–269. doi:10.1023/B:RIHE.0000019589.79439.6e

Tufekci, A., & Guzeldereli, E. A. (2018). Educational Data Mining In Distance education: A Sysyematic Literature Mapping Study. *International Women Online Journal of Distance Education, 7*(3), 61–73.

KEY TERMS AND DEFINITIONS

Educational Data Mining: An interdisciplinary field of study that is directly related to many fields such as computer science, statistics, mathematics, and data visualization.

Systematic Mapping: A method of describing the research available in a topic area.

Chapter 5

Management by Values in Educational Organizations:

A Case Study of a Technical University

Cemil Ceylan
Istanbul Technical University, Turkey

Büşra Aktaş
Istanbul Technical University, Turkey

ABSTRACT

Management by values (MBV) is a new form of management that creates a collective spirit and provides value-oriented work by combining people in an organization on common grounds. Even though existing research recognizes the importance of MBV in companies, less is known regarding MBV in educational organizations. This study provides an implementation of MBV at Istanbul Technical University (ITU) with an aim to increase the sense of belonging, create common organizational culture, reach targets, and perform higher so that the current engineering education provided is improved. The current situation of management at ITU was analyzed and an application method was developed, which was later implemented at Faculty of Management at ITU. Primary data used were collected from students, academicians, and managers of ITU via online surveys and face-to-face interviews. The results indicate that MBV, which has various benefits, should be used in educational organizations to build a more qualified educational environment, to raise more qualified students, and to have more satisfied members.

DOI: 10.4018/978-1-7998-2562-3.ch005

INTRODUCTION

Educational organizations are the institutions that form the foundations of an individual, provide connections between individuals, broaden the horizons of people, educate and develop the society, provide to gain perspective, and shape the foundations of thought. Rapid change and uncertainty all over the world with the development of digitalization affected educational institutions, which have critical importance. Education systems changed from document-based management to virtual learning environments (VLEs), which creates the need for interactions, cooperation, and common value systems (Schuster, 2017). This is invaluable to find ways to keep up with these conditions and to benefit from these developments to train engineers in the best way who are an important part of society and technology. In engineering education, it has become a necessity to use new management approaches in education to provide quality and contemporary education process that allows students and academicians to take the initiative considering that the people studying engineering are very fast-changing, rapidly developing, and productive minds. It is necessary to ensure that students and academicians are well understood, that their desires are taken into consideration by managers and that they play an active role in the management processes in order to understand the developments and tools that are emerging with digitalization, to keep up with the requirements of the changing world, and to implement them quickly and effectively. A difference can be made in engineering education by collecting students, academicians, and management in the common values to get the highest performance and yield. At this point, Management By Values (MBV) emerges and meets the requirements of current conditions.

MBV is a new form of management that creates a collective spirit and provides value-oriented work by combining people in common denominators in the organizations (Dolan and Richley, 2006; Akdemir, 2007). In organizations that are using MBV, they have an average of 5 to 7 specific values that unique for the organization, which are selected with respect to the values of every member in the organization. These values show what is important for the organization, guide them in all processes and decisions, and reflect the organization and its members. In this way, the opinions of every member are taken into consideration and management becomes a common process in which everyone participates instead of coming under the values of single persons. MBV is a new strategic leadership instrument that is a different way of sensing, processing, and applying information concerning behavioral sciences. MBV provides a strong structure to redesign the prevailing culture of organizations by shared values. The main objectives of MBV are simplifying, guiding, and securing commitment. Simplifying provides dealing with the complexity of organizations by harmonizing the changes in all standards. Guiding offers strategic management to be able to reach future targets of organizations. Securing commitment provides improvement of members' loyalty in daily works (Dolan and Garcia, 1999). Even all of the organizations have some level of values, MBV gives top priority to values in management and decision-making processes.

Value is the abstract measure to determine the significance of something and the response that something worth, based on the Turkish Language Society. Values are a set of principles that shape and control peoples' behaviors. Values affect peoples' decisions and beliefs about various actions. Therefore, values generate personal integrities and identities of people. These guide people to decide how important, useful, or accurate something is. Values have critical importance on the formation of individuals, organizations, and societies. These construct judgment of peoples. Values are general concepts that reflect the opinions, emotions, priorities, and objectives of the majority to ensure the continuity of the organizations. These are the basic standards that determine desirable or not-desirable, favorable or not-favorable, right or not-right of individuals, organizations, and societies. (Töremen and Akdemir, 2007).

The future of organizations depends on values that shape daily activities of their members from all levels. Organizations define and implement values according to their vision, missions, and expectations of their members to reach their targets. MBV presents a humanistic approach to the management of organizations that would provide the success of organizations in the current environment. It provides superior performance of members via regulating their actions (Jaakson, 2010) and enables leadership qualities to be noticed, which could contribute to creating higher success (Bukvić, 2016). MBV offers to increase the market potential and social status of organizations (Krzakıewıcz, 2012). It delivers a competitive advantage by creating unique organizational cultures that cannot be imitated by other organizations (Polat, 2006). MBV aims to reach top performance in organizations by motivating members, developing their capabilities and supporting synergies and innovations (Altınkurt and Yılmaz, 2010; Dolan and Richley, 2006).

The primary purpose of this study is to develop a model for implementing MBV, and apply the model in the Istanbul Technical University (ITU) as a case study with the aim to improve the teaching/learning system in engineering education. In the study, the concept of MBV and literature on MBV were reviewed and analyzed. Accordingly, a methodology was developed to implement MBV in an engineering education setting, where ITU was used as a case. During the implementation, it is aimed to support ITU to create a common organizational culture, to improve its education system, to reach its targets, and to increase its success. At the end of the implementation, critical points and insufficiencies of the method were detected, improved, and reported. The implementation included two phases, the quantitative and qualitative parts. In the quantitative section, online surveys were prepared and send to managers, academicians, and students in the management faculty of ITU via online tools. Thirty-two academicians and 96 students completed the questionnaires. Opinions about MBV, current and desired values in the ITU, sense of belonging to ITU, and demographic information of participants were collected. Common desired and insufficient values were determined based on the results. Regression analyses had done with collected data to find out affecting factors of sense of belonging to ITU. In the qualitative part, 15 academicians were interviewed in the semi-structured face to face forms based on the results of the quantitative section. Correspond actions to implement desired values and to figure out insufficient values were reported at the end of interviews. Finally, an application plan of MBV in the management faculty of ITU was constructed.

BACKGROUND

Value Systems

Value systems are divided into three categories to interpret behaviors such as economic-control values, ethical values, and emotional-development values. Economic-control values have been using for sustaining the organizations and bringing subsystems of organizations together. These values provide performance standards, effectiveness, profitability, and disciplines, such as organize, quality assurance, and accountancy. Ethical values have been used for defining members on how to behave in organizations while working and living, such as honesty, esteem, loyalty, and harmony. These values guide members for common beliefs. Emotional-development values have been used for providing organizations to construct and catch the opportunities and chances such as creativity, self-development, flexibility, adaptability, and self-reliance (Dolan and Richley, 2006; Dolan and Raich, 2013). Economic-control values, ethical values, and emotional-development values have been applied in different ratios concerning targets and members of organizations that provide organizational well-being. These three central values have three intersection points, such as survival, sensitivity-sustainability, and creativity-innovativeness. Survival exists between economic-control values and ethical values. Because these values provide managers easy decision-making processes by defining priorities and the critical points in organizational survival. Sensitivity-sustainability exists between moral values and emotional-development values. Because when the members of the organization are sensitive to the environment, they work for a better and sustainable world, which enables new developments. The creativity-innovativeness exists between the economic-control values and emotional-development values. Because organizations which support the passions of their members enable them to produce significant innovations. Such organizations own loyal and enthusiastic members who provide significant benefits to them. To conclude, organizations should define their priorities and requirements based on central values and their intersection points to be able to achieve their targets (Dolan and Raich, 2013).

In the 20th century, priorities and expectations of people had started to change, which caused a change in the value ranking. The benefits of organizations have not been the main objective of members' life after the late 20th century. Members do not want to postpone their desires or needs for the requirements of the organizations anymore. Members of organizations are more equipped, more educated, and more competent. Opportunities and possibilities of members increased, connecting them to the entire world, which cause to grow in members' requests. They want to see trust instead of control and to make their own choices in their ways. Therefore, the importance and effects of economic and control values had decreased, while the importance and impacts of emotional-development and ethical values have increased. Organizations should make the necessary changes to satisfy the request of their members in these circumstances. The way of satisfying the members of organizations is by providing individual supports for improving their potential and developing an organizational culture based on shared values (Anonymous, 2002; Dolan and Richley, 2006; Dolan and Raich, 2013).

Development of Management Tools

Different management styles such as Management by Instructions (MBI), Management by Objectives (MBO), and Management by Values (MBV) have been developed in time according to changing conditions, requirements, needs, and perceptions, respectively. In the MBI, rules, and materialism were in the foreground. Members whose work was limited to strict lines did not have the authority to take the initiative. Although it continued to be used between the 1920-1960s, MBI did not meet the needs of employees and created an insecure environment in the organizations. Finally, organizations have become unmanageable by MBI with increasing complexity and uncertainty all over the world. Organizations needed flexibility or adapting to changes and responding to the uncertainty. Therefore, MBO emerged, which provides the ability and possibility of own decision-making to managers and members by the 1960s. In the MBO, managers defined the activities that will enable organizations to achieve their goals. Managers were responsible for overseeing and ensuring that these activities are carried out. The managers set specific goals that were expected to be made by members at certain times. This way, organizations provided the members to work in a more flexible environment with higher motivation. However, even if the organizations had realistic and well-planned targets, and they could not reach their objectives because of the geometrical increase of uncertainty and unexpected conditions in the 21st century. Also, the hierarchic structure of organizations made members unmotivated. (Dolan and Garcia, 1999; Anonymous, 2002; Dolan and Richley, 2006; Dolan and Raich, 2013).

In the 21st century, organizations became more dynamic and chaotic systems because of the rapid changes, complexity, and globalization. According to the literature, predicting and knowing what will happen depend on understanding the values of living systems. These values are the core concepts to interpret the behaviors and responses of individuals and organizations. Therefore, MBV emerged as a response to these circumstances in the world. MBV means defining common values based on organizations' targets, brand and past image, requirements, and opinions of their members. These shared values guide managers and members and define priorities in their decision-making processes, general issues, and daily work. MBV overcomes some of the disadvantages of the previous models and take advantage of the current chaotic environment. Because when organizations are capable of defining their common value systems, they can satisfy their members' requirements, manage chaos, and be sustainable. Even these chaotic environments make members advance in their works and relations to be useful for organizations. To conclude, MBV became the main part of the organizations' management systems, which provides strategic leadership and advantages with high potential (Dolan and Garcia, 1999; Anonymous, 2002; Dolan and Richley, 2006; Dolan and Raich, 2013).

Basically, MBV exists to meet the four primary requirements, which causes complexity and uncertainty in the current environment. The first requirement is the need for quality/customer orientation. Because the competitive and complex environment made mass production and market segmentations insufficient, individualization of products or services became mandatory. The second requirement is the need for professionalism, autonomy, and responsibility. With the advance of technology, the demand for quality and customization increased. Therefore, organizations expected more competency, capability, creativity, loyalty, and flexibility from their members. The third reason is the need for supervisors/managers to evolve into leaders/facilitators. while supervisors use instructions to manage the organizations, leaders use values. Leaders can keep the organizations together, inspire the members, ensure the efficiency-effectiveness of work-done, and guide the processes and members. The fourth and final reason is the need for a flatter and more agile organizational structure. Because inflexible, hierarchical, rigid, and bureaucratic structures make organizations less efficient. These structures are not sufficient for being sustainable and prosperous in the current environment. Flexible and horizontal structures rather than pyramid structures are able to overcome uncertainties in organizations. More detailed advantages and differences between the MBV and the other management styles about various areas are seen in Table 1. To conclude, MBV provides these requirements, which are individualization, creativity, loyalty, flexibility, inspiration, efficiency, and horizontal structure briefly (Dolan and Garcia, 2002).

Table 1. Differences between MBI, MBO, and MBV (Dolan and Garcia, 2002).

	MBI	MBO	MBV
Preferable situation for application	Routine or emergencies	Moderate complexity	Need for creativity in the solution of complex problems
Average level of professionalism of members of the organization	Basic level of education (management of operatives)	Moderate to average professionalism (management of employees)	High level of average Professionalism (management of professionals)
Type of leadership	Traditional	Allocator of resources	Transformational
Image of customer	User-buyer	User customer	Customer with judgment and freedom of choice
Type of product market	Monopolist Standardized	Segmented	Highly diversified and dynamic
Type of organizational structure	Pyramidal with many levels	Pyramidal with few levels	Networks, functional alliances, project team structures
Need for the tolerance of ambiguity	Low	Medium	High
Need for autonomy and responsibility	Low	Medium	High
Stability of environment	Stable environment	Moderately changeable environment	Very dynamic, changeable environment
Social organization	Capitalist-industrial	Capitalist post- industrial	Post-capitalist
Philosophy of control	"Top-down" control and supervision	Control and stimulus of professional performance	Encouragement of self-supervision by each individual
Purpose of the organization	Maintenance of production	Optimization of results	Continuous improvement of processes
Research of strategic vision	Short term	Medium-term	Long term
Basic cultural values	Quantitative production, loyalty, conformity and discipline	Rationalization, Motivation, Efficiency, Measurement of results	Developing participation, continuous learning, creativity, mutual trust, commitment

Concept of Management by Values

MBV is a new method that combines strategic management and human resources management (Rokeach, 1983). Human resource management is one of the biggest challenges of managers in organizations. Because people are complex, and everyone has their own values (Dolan and Altman, 2012). According to Krzakiewicz (2012), motivations of people are provided via knowledge, empowerment, and involvement. When the values of organizations and members are clear, the motivation and the efficiency of the members increase. When a connection between members and management is present, members' creativity and willingness to self-development increase. Members engage in organizations via common values (Krzakiewicz, 2012). Therefore, MBV is of importance for organizations that want to develop an organizational culture. Organizational culture is formed by defining common values based on the aims and principles of the organization. Managers have more significant responsibility and effect in the application of corporate culture because of their authority. They should make sure that shared values are used at all levels and processes of the organization. Also, managers should continuously analyze and improve the strengths and weaknesses of the organization to handle complexity, uncertainty, and change. When the variations exist, organizational culture should be redesigned to adopt and observe the changes. Management of change can only be achieved through efficient leadership.

The main difference between successful organizations and the others is that successful organizations have organizational culture (Peter and Waterman, 1982). Organizational culture provides strategic and competitive advantages such as economic, social, political, technological, and organizational. A clear definition and execution of corporate culture make members creative, hardworking, and loyal to organizations. Members have the opportunity to be a part of active management by taking incentives based on corporate cultures. In the organizations which apply MBV successfully, members know their organizational values, personal values, and their meaning for their organizations. Therefore, their motivation and performances improve. Organizational culture guides managers and the members of the organizations in both daily activities and critical issues. These provide to see priorities and the crucial point in the organizations easily (Anonymous, 2002; Dolan and Richley, 2006; Dolan and Raich, 2013).

MBV makes it possible to design and redesign organizational cultures. According to Schein (1992), the first level is the shared values, assumptions, and beliefs which are constructing the organizational cultures. These are using to manage daily works and providing loyalty to organizations. The second level is cultural artifacts such as procedures, rituals, logos, advertising, sponsorships, and behaviors (Schein, 1992). Consistency should be provided between these two levels in all statuses to construct a strong culture (Dolan and Richley, 2006). While the organizational culture is being built, the value systems of every member must be taken into consideration to be able to provide effectiveness. Value systems should be constructed based on brand and past image of the organization. Values should be permanent and should not change based on different managers. Values must be chosen that cannot be able to imitate by other organizations for creating a competitive advantage (Polat,

2006). Expectancies and requirements of current and potential members should be considered. Organizations should make sure that new members assimilate common values. Finally, all decisions and behaviors must be based on these values. Managers should be objective by taking their decisions based on shared values instead of their values.

Management by Values in the Literature

Beck (2014) reports that 75% of large companies in Europe, 66% of large companies in the UK, 50% of large companies in France, and 33% of large companies in Norway and Sweden are applying MBV. This management tool has shown success in organizations such as Coca Cola, and Briggs & Stratton. Therefore, various studies on MBV in different kinds of organizations can be found in the literature that is studying the benefits of MBV and different claims about MBV. Some of these benefits and applications are as follows: the existing organizational values will make organizations profitable (Blanchard, 2014); MBV creates, improves, and sustains productive values (Harung and Dahl, 1995); corporate appraisals should be consistent with daily practices to prevent insecurity, estrangement, and indiscipline (Kerwin, MacLein, and Bell-Laroche, 2014); the culture of managers and culture of organizations are interacting with each other (Demirtaş and Kılıç, 2015); organizational culture provides to increase consistency, performance, loyalty and provides to reach targets in organizations; success increases when the managers balanced their values with organizational values (Doğan, 2016); every organization that cares about its image must have clearly defined mission, vision, and organizational values (Šundov and Dulčić, 2017). Also, there are studies reporting the success of MBV in different industries, such as the automotive industry (Beck, 2014).

Various but similar results were reached with these studies and explained as follows. Organizations are collaborations where shared values and actions are existing. Every organization has specific values that refer to its mission and vision; even the common culture did not define. Also, the reflects of MBV did not exist in the organization; perceptions for MBV are positive (Demirtaş and Kılıç, 2015; Doğan, 2016; Šundov and Dulčić, 2017).

The process of MBV in organizations begins with the participation of all members. To apply MBV, all members of the organization should show an effort to changing existed values to common values. Firstly, existed values in organizations should be discovered. Secondly, organizations should define their mission, vision, and common values. Thirdly, the implementation of MBV should be done with the guidance of managers and common values. As a final, the critical point changes from 'doing thing rights' to 'doing right things,' and from 'efficiency' to 'effectiveness' with applying MBV in the organizations (Harung and Dahl, 1995; Blanchard, 2014; Beck, 2014; Doğan, 2016).

Common values should be decided in collaboration with members via discussing which provide to adopt values in all levels of the organization. Values should be defined as answerable and rational to gain the most efficiency and consistency. Personal values and organizational values should be combined to provide expressive and meaningful organizational lives. Values should be connected and consistent with daily activities and strategies. These values should reflect in all levels of management. Organizational culture should be constructed and reflected in daily works to reach the targets of the organization (Blanchard, 2014; Kerwin, MacLein and Bell-Laroche, 2014; Demirtaş and Kılıç, 2015; Doğan, 2016; Šundov and Dulčić, 2017).

MBV increases the importance of human resources and efficiency because the most important and precious resources of organizations are humans. Organizations should hire people which compatible with their organizational culture. Because when people are compatible with organizations, their tenure and self-control increases (Harung and Dahl, 1995).

Managers have a significant role in the application of MBV who can launch and guide MBV with human resources management. Managers, which can be affected by their values, should be flexible in applying common values. They should manage the image of the organizations and should create differences by constructing common values. Managers should trust employees, which provides trust and self-development to members. It provides joy-motivation instead of fear-motivation. Thus, managers can ensure the success and fulfill expectations of all members. Satisfied members are the essential power of the organizations. When the employees are satisfied, their tenure increases, which provides constant culture. This way, organizations can invest in the development of members instead of hiring costs (Harung and Dahl, 1995; Kerwin, MacLein, and Bell-Laroche, 2014; Šundov and Dulčić, 2017).

To conclude, MBV makes organizations more successful and prestigious that provide financial support and loyalty, which in turn increases the connectedness between all levels of the organization. MBV provides benefits in performance management, budget planning, and organizational structure. MBV provides flexibility, profitability, and stronger management (Harung and Dahl, 1995; Beck, 2014; Kerwin, MacLein, and Bell-Laroche, 2014).

Management by Values in Educational Organizations

The current education system was built based on the industrial revolution, which caused the authority and administration of single persons. Therefore, the priorities and preferences of industrial societies exist in the background of today's' education philosophy (Töremen and Akdemir, 2007). Common value systems could not be provided or improved in these circumstances because of the central dependencies. However, when the organizational cultures based on shared value systems do not exist, managers apply their actions concerning their own value systems, which affect organizations negatively. When the requirements, outlooks, and values of members in education organizations are not satisfied, members lose their motivation, loyalty, benefit, and commitment. Educational organizations, which are critical for society, should not be managed by a single persons' value systems (Töremen and Akdemir,2007).

In the 21st century, social sciences were shaping the environment. Human management became one of the biggest challenges of management, especially in educational organizations. MBV became a necessity in the educational organizations with the digitalization, which caused the diversification of education systems, expectations of people, education tools, and dissemination rate of developments. Multiculturalism, participation, transparency, and willingness became foreground issues in the field of education. Members start to expect educational organizations to make developments, innovations and take responsibilities (Töremen and Akdemir, 2007). Thus, managers and academicians of these organizations should be leaders instead of supervisors, should be looking for developments and innovations, should enable members to participate in management (Töremen and Akdemir, 2007). Value systems of every member must be taken into consideration to be able to provide effective, successful, and beneficial education (Dolan and Altman, 2012). Value systems of educational organizations should be constructed based on common expectancies and requirements of their current and potential members (Walters, 2002).

MBV provides significant and various benefits to educational organizations. It provides strategic choices to develop the individual and professional potential for each member of the educational organizations. It supports changes, developments, and innovations. This new management tool provides organizational success via defining organizational culture (Doğan, 2016). Organizational culture guides to managers, academicians, and students in both critical decisions and daily issues. Organizational culture makes academicians motivated, and motivated academicians make educational organizations successful. Organizational cultures provide a vivid and meaningful education life to members (Doğan, 2016). It provides commitment, high-level education, more successful students. To conclude, MBV should be applied in engineering education to take its advantages in current situations.

In the literature, there are various studies about the MBV in the educational organizations. Research processes showed that managers, academicians, and students support MBV in educational organizations. Application processes showed that similar methods were applied in different studies. Firstly, MBV were researched in detail with the experts of the MBV (Demirtaş and Kılıç, 2015). Studies were conducted via qualitative method or mixed-method, which includes both quantitative and qualitative approach (Kerwin, MacLein and Bell-Laroche, 2014; Demirtaş and Kılıç, 2015; Doğan, 2016). A mixed-method was considered a better option to make analyzes (Creswell, 2013). The variety was preferred while constructing the study groups. In the quantitative parts, demographic information was collected. Managers/ assistant directors of different middle schools/high schools or teachers of different elementary schools were selected to discover different perspectives. Gender, age, branch, education degree, tenure, graduation faculty, number of teachers in the school, and school type of participants considered independent variables to discover whether there is a relation between values and the variables. After determining the questions based on researches, semi-structured interviews and observations were done with participants. The audio recording was used to save data (Demirtaş and Kılıç, 2015; Doğan, 2016; Šundov and Dulčić, 2017; Kerwin, MacLein and Bell-Laroche, 2014). In the qualitative part, semi-structured interviews were done to analyze the concept and effects of MBV and the opinions of participants about MBV. The data analyzed with ANOVA and similar analysis (Doğan, 2016; Beck, 2014). The results were reported based on the studies.

Applying Management by Values In A University

The main focus of the chapter is applying MBV in the Istanbul Technical University (ITU) to improve the teaching system in engineering education. With this purpose, the current situation of management in the ITU was analyzed, an application method was developed and implemented in the management faculty of ITU. ITU is one of the oldest technical universities in the world, which founded in 1773 and nearing its 250th anniversary. ITU offers graduate and doctorate programs in 6 institutes and 39 undergraduate programs in 13 faculties in 5 different campuses. The reason why we chose ITU is that ITU is an institution branded with history, science, technology, art, and sports, which is identified with the fields of engineering and architecture. ITU is a long-established and leading university that continually develops and renews (itu.edu.tr). In cooperation with students at ITU, we aimed to support the creation of a lasting and common culture through the implementation of MBV by academicians and administrators. Also, we intended to increase awareness of authorities about MBV and provides them a guide in the process of application. Because authorities, which are managers and academicians, have critical roles in the application of MBV.

The study utilizes a mixed method, which includes both quantitative and qualitative data in 14 weeks. The mixed-method is considered as optimal because the method provides to make better analyzes (Doğan, 2016 and Creswell, 2013). Maximizing the variety of participants was aimed to reach more reliable results, and writing had used to save data. In the quantitative part, online surveys had done with students, academicians, and managers after defining and introducing the concept of MBV to participants briefly. Two different online surveys had prepared separately for students and academicians/managers of the industrial engineering department, management engineering department, and economics department in the management faculty. 96 students and 32 academicians/managers had answered the surveys. In the online surveys, firstly, the sense of belonging to the university was analyzed based on the variables. The variables of students were gender, department, and grade. The variables of academicians were gender, task, department, tenure, and title. Regression analyses was applied to the collected data to discover the relation between the sense of belonging and demographic information. Secondly, the opinions of participants about MBV were analyzed. Participants were asked to choose whether they think that MBV is beneficial and whether they would support MBV. It is aimed to find the agreement level. Thirdly, current values and desired values in the faculty were found. Sixteen values based on literature researches were presented to the participant to see whether these values are using in the faculty and which values they want to be used in the faculty. Form of online surveys have shown in Appendix A and Appendix B. In the qualitative part, opinions of academicians/managers were collected with semi-structured face-to-face interviews, which were prepared based on the results of the regression analyses and statistical data. These interviews had done with 15 academicians from 5 different degrees. Three academicians were chosen for interviews from each degree. These degrees are Professor Doctor, Associate Professor, Doctor Lecturer, Lecturer Doctor, and Research Assistant. Questions of interviews have shown in Appendix C.

In the quantitative part, the aim of online surveys was introducing MBV, collecting statistical information about the affecting factors of MBV, agreement degree to MBV, current and desired values, and sense of belonging in the Management Faculty of ITU with a point of view of students and academicians/managers to analyze the existing structure. In the qualitative part, the aims of interviews were discussing MBV, current and past situation of the ITU, desired and insufficient values in the faculty, application process of MBV, reasons of insufficiencies, and requirements by the results of online surveys. As a result of this study, we aimed to analyze the current situation at the Management Faculty of ITU, to find common values for creating organizational culture, and to present a revised plan that will show the critical and essential points to be able to implement MBV successfully.

Findings

The two-stage study showed different results with the analysis of the collected data.

Table 2. Opinions of students

Variables		f	%
Opinions about MBV	Beneficial	80	83,3
	Non-Beneficial	16	16,7
Belonging rate to ITU	1	0	0
	2	22	22,9
	3	46	47,9
	4	23	24
	5	5	5,2

Interpretation of Quantitative Part (Regression Analyzes)

In the quantitative section, two online surveys were done for students and academicians/managers separately. For both students and academicians, gender, department, class, task, tenure, and title of participants were asked to find out whether these factors affect opinions about MBV and sense of belonging rate to ITU. Current values in ITU were asked to find out whether ITU has clear existing values. Desired values in ITU were asked to find out common desired organizational values. Recommendations were asked to provide space for any suggestions.

Opinions of students and academicians were collected and shown in Table 2 and Table 3. Based on these data, the first result is that 83,3% of students and 93,8% of academicians agree that MBV will support ITU. They agree that the MBV is beneficial for ITU, which is important for creating awareness about the MBV and encouragement to apply. This high level of support, which can provide to make the study effective and permanent, is inspiring and promising. The second result is that the average sense of belonging rate to ITU is 2.97/5 at students and 3.25/5 at academicians, which should be improved. It shows that MBV, which provides to develop a sense of belonging to organizations by constructing common values and collective soul, is necessary for ITU.

The demographic data of students and academicians were collected and shown in Table 4 and Table 5. Multiple linear regression analysis is applied to the data collected from students and academicians/managers separately. Multiple regression analysis is "a general statistical technique used to analyze the relationship between a single dependent variable and several independent variables." Backward elimination was used for regression analyses, which were defined as "method of selecting variables for inclusion in the regression model that starts with all independent variables in the model and then eliminates those variables that do not make a significant contribution to prediction." (Hair et al., 1995). Dummy variables, which means "the independent

Table 3. Views of academicians

Variables		f	%
Opinions about MBV	**Beneficial**	**30**	**93,8**
	Non-Beneficial	2	6,3
Belonging rate to ITU	1	0	0
	2	7	21,9
	3	3	34,4
	4	9	28,1
	5	5	15,6

Table 4. Demographic information of students.

Variables		F	%
Participation Rate	Industrial Engineering Department	39	30,2
	Management Engineering Department	28	29,2
	Economics Department	29	30,2
Gender	Female	52	54,2
	Male	44	45,8
Class	1	20	20,8
	2	23	24
	3	22	22,9
	4	31	32,3

variable used to account for the effect that different levels of a nonmetric variable have in predicting the criterion variable," were used to express nominal values as ordinal values in the regression analyses. It is aimed to find a valid model and significant values that explain the collected data.

For students, independent variables are gender, department, and class, while the dependent variable is a sense of belonging rate to ITU. Dummy variables are used to express gender, department, and class, as shown in Table 6 and Table 7. Regression analyses was done with demographic information of students. Result tables of regression analyses shown in Appendix D. The best model are decided via adjusted R square value based on the table of Model Summary. The best model is model 4, which has the biggest adjusted R square value as 0.489. Model 4 is valid

Table 5. Demographic data of academicians/managers

Variables		F	%
Participation Rate	Industrial Engineering Department	15	46,9
	Management Engineering Department	16	50
	Economics Department	1	3,1
Gender	Female	20	62,5
	Male	12	37,5
Professional Seniority	1-5 years	2	6,3
	6-10 years	13	40,6
	11-15 years	4	12,5
	16 years and above	13	40,6
Task	Academician	27	84,4
	Manager and Academician	5	15,6
Seniority of Managers	1-3 years	5	50
	4-6 years	3	30
	10 years and above	2	20
Degree	Research Assistant	12	37,5
	Lecturer Doctor	1	3,1
	Doctor Lecturer	5	15,6
	Associate Professor	7	21,9
	Professor Doctor	7	21,9

Table 6. Dummy variables of gender, task, class, and degree

Gender	Variable	Task	Variable	Class	Variable	Degree	Variable
Female	1	**Academician**	0	**1st class**	1	**Prof.**	5
Male	0	**Manager**	1	**2nd class**	2	**Ass.Prof.**	4
				3rd class	3	**Doc.Lec.**	3
				4th class	4	**Lec.Doc.**	2
						Res.Ass.	1

Table 7.Dummy variables of departments

Departments	Variables		
	IndEng	ManEng	Eco
Industrial Engineering Department	1	0	0
Management Engineering Department	0	1	0
Economics Department	0	0	1

because its significance value is less than 0,05, as shown in the table of ANOVA. The model showed that only significant value is a class of students that affect the sense of belonging. The significance value of the class of students is less than 0.05 as shown in the table of Coefficients. Therefore this value is valid. As a result, the regression analysis showed that while the class of students is increasing, the belonging rate to ITU is decreasing. The linear regression model of sense of belonging rate to ITU is as follows based on Model 4.

Sense of belonging rate to ITU = 4.615 - 2.53*Class of students.

For academicians, independent variables are gender, department, task, tenure of academicians, tenure of academicians with the managerial task, and title. The dependent variable is a sense of belonging rate to ITU. Dummy variables were used to express gender, department, task, and title, as shown above. Regression analyses had done with demographic information of academicians. Result tables of regression analyses shown in Appendix E. The best model is decided via adjusted R square value based on the table of Model Summary. The best model is Model 5, which has the biggest adjusted R square value as 0.268. Model 5 is valid because the significance value of Model 5 is 0.008, which is less than 0.05, as shown in the table of ANOVA. The model showed that department, task, and degree are significant values because the significance values of these variables are less than 0,05, as shown in the table of coefficients. Therefore, this value is valid. As a result, while the academicians/managers are in the department of management engineering and while the academicians have a managerial task, the sense of belonging to ITU is increasing. However, while the title of academicians is increasing, the sense of belonging rate to ITU is decreasing. The linear regression model of sense of belonging rate to ITU is as follows based on Model 5.

Figure 1. Current values relating to students.

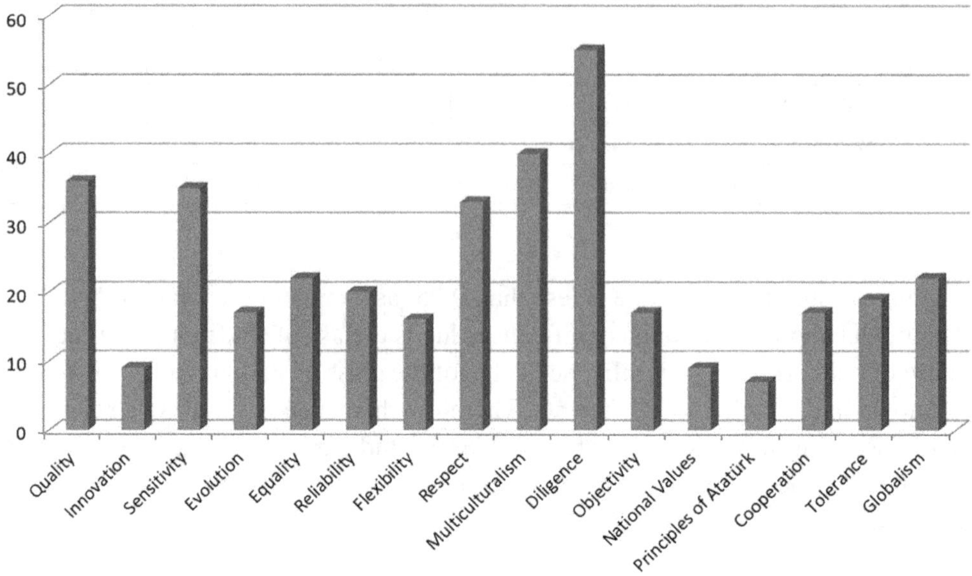

Sense of belonging rate to ITU = 3,77 + 1.982*ManEng + 0.885*ManAca -2.53*Title.

To conclude, the common factor for students and academicians that affect the sense of belonging rate to ITU is time spent in the university. As long as the class of students and the title of academicians are increasing, the sense of belonging to ITU is decreasing.

Interpretation of quantitative part (Current and desired values)

Students and academicians/managers were asked to choose existing values in the ITU. For both academicians and students, diligence and quality are currently existing values, while equality and objectivity are insufficient values based on the online survey results. Current values concerning students and academicians were shown in figures 1 and 2.

Figure 2. Current values relating to academicians/managers

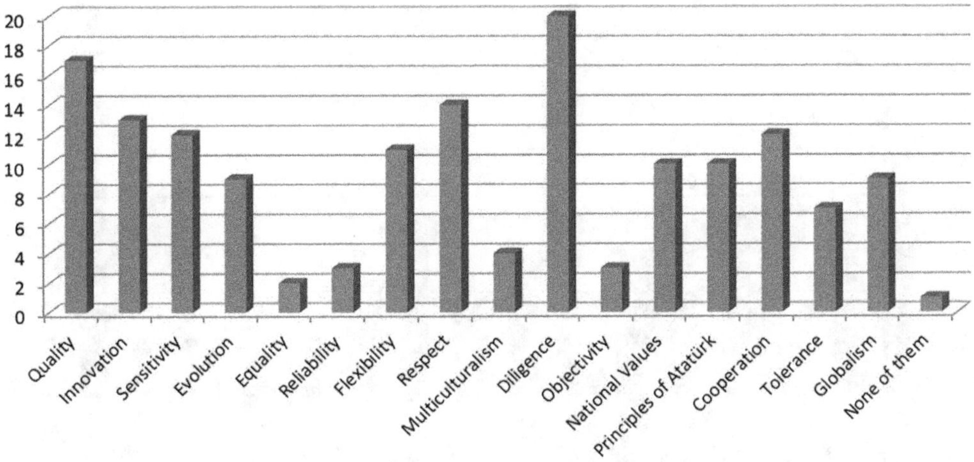

Students and academicians were also asked to choose four values which they desire to represent ITU. Based on online survey results, desired values for students and academicians were shown in figures 4 and 5. For academicians, innovation, equality, objectivity, and globalism are the four most desired values. For students, quality, innovation, diligence, and globalism are most four most desired values. To conclude, common insufficient values, which should be improved, are objectivity and equality. Four most desired values are quality, innovation, diligence, and globalism based on the total rate of answers. Definitions of desired and insufficient values are as follow based on the Cambridge Dictionary. Quality means high standards in both technical and social issues. Innovativeness means using new methods and ideas in the education and management system of ITU. Globalism means including or affecting the whole world. It means connection and accordance with ITU to the world in social, educational, and technological aspects. Diligence means the quality of working carefully and with a lot of effort. Equality means to behave and judge the members equally without considering age, belief, gender, nation, department, etc. Objectivity means the fact of being based on facts and not influenced by personal beliefs or feelings.

Figure 3. Desired values concerning students

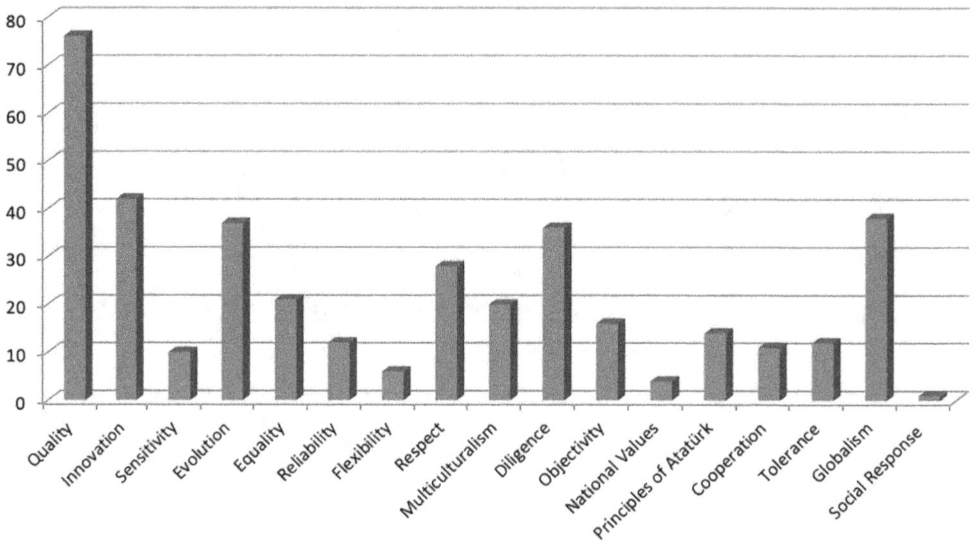

Figure 4. Desired values concerning academicians/managers

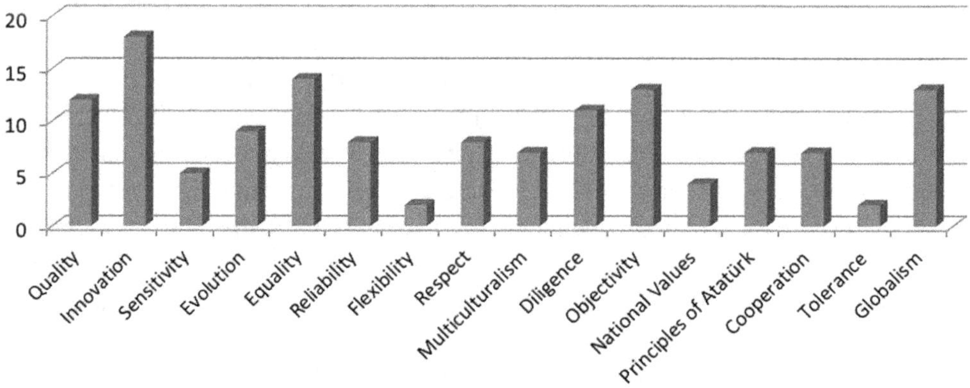

Interpretation of the Qualitative Part

The qualitative part, which consists of four questions, was prepared based on the results of the quantitative section. Each item was asked after a brief explanation based on the results that we obtained. Description of the first question was "The MBI and MBO are replaced by MBV in all over the world due to a rapid and uncertain environment. Organizations are aiming to increase their success and loyalty by constructing common values and taking their decisions for these common values.

Based on the survey result, more than 90% of academicians and %80 of students think MBV is beneficial for ITU." The first question, which is '*What does MBV means to you?*', were asked to find out opinions, knowledge level, and awareness level of academicians about the MBV. Answers of the first question showed that 46% of academicians have an idea about MBV. Only 33% of academicians could define the MBV clearly. It indicates that MBV has not known or spread within academicians of ITU. Although MBV has not known sufficiently, there was no negative approach to the method.

Explanation of the second question was, "The results of surveys show that current values are existing less than desired rates. Some of the participants specified that values are not using anymore in management of ITU." The second question, which is '*Do you think any attention is paid to values in the management of ITU? Do you think there is a difference between the past and present situation of management in ITU?*', were asked to find out whether the MBV exists in ITU in any level, and whether the difference is existing in ITU in management. Answers of the second question showed that only 20% of academicians said that values are given importance in the management of ITU. 66% of academicians have sufficient experience in ITU to compare past and current situation of management in ITU due to their tenure. Only one of these academicians said that conditions are better concerning history. Other academicians said that values were given importance in management. However, values are not given importance in ITU currently. The management uses its values, which are more subjective and unfair. They defended that ITU had a culture which is no longer existing. Values are changing with the change of management, which is a fault.

Explanation of the third question was, "The results of surveys show that most desired values in the ITU are quality, innovativeness, globalism, and diligence. Insufficient values are equality and objectivity. " The third question, which is '*What do you think about the most desired and insufficient values in the management of ITU? Do you think the desired values are logical? What are the reasons for the insufficient values?*' were asked to find out whether desired values are applicable, and to find out the opinions about lack of the values. Answers of the third question showed that 93% of academicians considered equality and objectivity as insufficient. All academicians said that globalism, innovation, diligence, and quality are logical and applicable. Correspond actions to desired and insufficient values are listed in the title 'Solutions and Recommendations.'

The final and fourth question, which is '*How can we apply management by values in ITU? Do you think the administration will support the MBV? What can academicians do to disseminate common values to students?*', were asked to actualizing MBV, announcing common values, and find out the application method. The explanation of the fourth question was, "Most important persons in the application process of MBV are managers of the organizations. Common values can be applied, disseminated, and clarified successfully and permanently with the collaboration of managers. Based on surveys, the common factor in students and academicians, which affecting the belonging rate, is the degree. When the class of students is increasing, and the degree of academicians is increasing, belonging rate to ITU is decreasing." Answers of fourth questions showed that 80% of academicians admit the accuracy of the relation between the sense of belonging rate to ITU and the degree of students/academicians. The reasons for this relation explained as follows. For academicians, they are not given sufficient importance. When the experiences of academicians increase, more opportunities expected from ITU, but expectations are not met. As the title of academicians increases, their loyalty to ITU decreases. Equality and objectivity are lack. For students, they consider ITU as a perfect target, but ITU is not meet the expectations. The academic benefit is not enough. Management faculty cannot provide campus life, which creates social and physical inadequacies. After a while, students start to find ITU deficient because they have begun to compare ITU with different universities. Academicians made many application proposals which used in the application plan.

SOLUTIONS AND RECOMMENDATIONS

Even the ITU is one of the most successful technical universities in the world, it has some insufficiencies and must be improved points. The fundamental problems of the ITU are lack of digitalization and failure to meet the expectation of members. This situation can be solved and improved by applying MBV in the organization. MBV should be applied to create a common culture that can increase the sense of belonging to ITU via creating a humble spirit. Because success and development cannot be achieved without shared commitment (Bell and Harrison, 2018). To make a successful application of MBV, values that will be implemented should be selected so carefully.

Insufficient values based on the study were defined as equality and objectivity. Correspond actions for equality and objectivity are as follows. The value of transparency should be used in all of the processes to provide equal and objective authority. Transparency of processes such as grading of student's exams or projects, recruitment and advancing of academicians, or budget allocation of projects/studies can be provided with digitalization. As an example, financial resources or promotions of academicians can be defined based on their score, which carried out over a transparent digital system and determined by factors such as many publications, course hours, project numbers, or student evaluations. Such a system can prevent discrimination and disbelief. The decision-making mechanism of all systems should be designed according to common values. The decision-making process must be spread to departments instead of collecting in the deanery. Belonging to ITU must come forefront instead of belonging to groups. Assistants must be able to work with all academicians instead of depending only on their instructors.

Most desired values based on the study were defined as quality, innovation, diligence, and globalism. These values should be used in the ITU by setting corresponding actions. Based on the answers of academicians and researches, correspond actions for quality are as follows. Physical conditions of the faculty must be improved, and facilities such as computer laboratories, projection machines in classes, and libraries must be improved. Adequate training to students in different areas must be presented, such as coding, data analysis, or virtual reality to use the advantages of digitalization. To earning higher quality academicians to ITU, opportunities presented to academicians must be improved. Extra support must be given to academicians for publishing more articles and studying more projects. There should be an encouragement to support academicians who teach better. Social activities, such as social responsibility projects, should be done to improve the humanitarian of students. Academicians must increase their connection and support to students. Correspond actions for innovation are as follows. Firstly, for ITU to be innovative, individuals should be creative in their minds. Innovation should be supported by management, and old methods must be replaced with new methods without losing the benefits of a classic. Original and different contents must be produced. Social, organizational, educational innovations must be increased.

To conclude, values of quality and innovation should be used in the ITU via digital technologies to improve education systems. On the other hand, diligence represents ITU, even in the current situation. This value should always be used in the ITU to achieve more and more success. Correspond actions for diligence are as follows. Individual must take responsibility to improve ITU and their self. Motivation provider measures should be taken, and an incentive system must be built to encouraging the studies. Academicians and students must have a minimum limit for their studies, such as projects and articles. In the current conditions of the world, globalism is not a choice but an obligation. Correspond actions for globalism are as follows. Firstly, for ITU to be global, individuals must be universal in their minds. If the individuals give importance to globalism, they use their resources and make an effort in this direction. ITU must adapt to international standards, and new course contents should be developed for global developments. ITU should establish and improve its relations with international organizations without intermediaries. International publications, international teachers, and international contracted programs should be increased. Financial support for international studies should be improved because globalism cannot develop due to the cost factor. To conclude, ITU can offer many more opportunities to members and can achieve more significant successes with solving these problems by applying MBV with the proper values.

Application Plan of MBV in Educational Organizations Based on the Study

Application plan of MBV in the educational organizations constructed based on the study. The method consists of five steps. The first step is the preliminary preparation and research process. A core team, which will be the manager and advisor of the application process, should be built. This team should be consist of academicians and managers from different departments and different titles. Because the study showed that departments and titles of academicians affect a sense of belonging to the university. The members of the core team should be selected from the ones who know the management and MBV. The core team should research the concept of MBV with details. After definitional and conceptual researches are done, application examples of MBV should be researched. Different educational organizations which succeed the built and maintain MBV with common culture and value system should be analyzed. Academicians and managers of these organizations should be consulted. The second step is reporting and sharing. The researches, which had done by the core team, should be published and shared with all academicians and the managers in the organization by e-mails, meetings, and conferences. The requirement and benefits of MBV should be known and understood by all the academicians and managers. When they can dominate the MBV, students should be involved in studying. The report of the study, importance, benefits, and examples of MBV should be shared and explained to students in lessons and conferences. Mainly, MBV should be explained and shared in the lessons by academicians to students in a specific period. Because the study showed that the average sense of belonging to the university is lower in the students for academicians. These conditions could be improved by including students in the processes. The third step is creating motivation. The study should be progressed simultaneously and transparently between academicians, managers, and students. The opportunity to participate in the study should be presented to students to develop the organization together. Motivation to create a common culture must be ensured with a joint effort. The fourth step is finding common values and corresponding actions. Opinions, recommendations, and requirements in the organization should be analyzed. Current values, desired values, and insufficient values in the organization should be found. Common values should be decided and defined based on these researches. After identified the common values, the corresponding actions that need to be done to implement these values should be ordered separately. The fifth and final step is making MBV permanent in the organization. An incentive system should be established to encourage and ensure the continuity of implementation until the values are fully established in the organization. Values should be announced to the public via social media tools of the organization. Values should be reported and explained since the orientation to students, academicians, and managers. Appraisals should be seen in the aisles of the organization. Because the study showed that even the organization had written values, members are not aware of them if the values are not using or announcing. To conclude, new members should assimilate the values for harmonizing them with organizational culture.

In the application of MBV, critical points exist which need attention. The identified common values, which creates organizational culture, should be clarified and widespread to all of the members. These values should overlap past and brand image of the organization and should be specific to the organization. Everyone should understand the same meanings from the values. Values should be permanent and should not change based on different managers. All decisions and behaviors should be done based on these values. Managers should be objective by taking their decision based on common values instead of their values. The management process should be spread between lower and upper levels. Organizational culture should have cared for more than personal values by students, academicians, and managers.

FUTURE RESEARCH DIRECTIONS

Management by Instructions (MBI) and Management by Objectives (MBO) replaced by Management by Values (MBV) with the rapid change and uncertainty in the world due to the effects of digitalization. MBV focuses on the general image and cultural preferences of organizations instead of details of processes. Instead of planning and controlling every business, the organizations can focus on their culture, policy, developments, and strategy with the MBV. When the studies showed that specific applications or programs are not the critical points to manage digitalization, but the management of the development's functions are critical; once again, the importance of MBV is emerged (Benedek and Molnar, 2015). MBV provides to adapt new trends and new tools of digitalization via flexible and value-based management.

In the current situation, the management and decision-making processes have become much more horizontal because people want a much more flexible environment. In the future of the education system, a digital scoring system can be established for students to get the grades they deserve, not only through exams but also with their different abilities, projects, or works. Also, the online exams, lessons, projects, etc. can take the place of the classical system. In addition to basic engineering knowledge, courses in industry 4.0, such as augmented reality and coding, can be given more emphasis in the curriculum.

In the application of MBV, digital tools provide to find common values for all members of organizations. Also, these tools provide to release common values to the members and the public in an easier and faster way. The implementation of MBV is supported and accepted by everyone because MBV provides to satisfy the requirements of the time and demands of members. A case study about the application of MBV in the management faculty of ITU had done with the 128 participants and limited numbers of variables. In the future, the study could be released to all faculties of the ITU, numbers of participants and variables could be increased to reach more general results.

CONCLUSION

In this study, the concept, benefits, and necessity of MBV have been tried to be considered and understood. Studies in the literature and the application examples about MBV have been analyzed. Based on the literature research, a limited number of studies have been conducted on MBV in educational organizations. This research and the study showed that MBV had not given sufficient importance in educational organizations. However, in the new age, MBV is existing, spreading, and applying more and more in organizations because of the changes, complexity, and uncertainty in the whole world. The reason for these situations is digitalization, which provides to spread news, trends, or inventions all over the world very quickly and instantly. Digitalization reflects in all of the processes in organizations such as decision-making, promotion, R&D, or marketing. Thus, the demands of people and organizations have been changing from basic needs to self-actualization, which can be satisfied by given attention to social and behavioral sciences such as MBV. Organizations can harmonize these circumstances via MBV, which is a new form of management. MBV provides to define priorities, values, and aims of the organizations. MBV provides organizational culture, better management, higher standards, loyalty, a sense of belonging, satisfaction, profitability, and a better brand image. MBV provides to meet the expectations of members in the organizations. Therefore, this study had done to bring the advantages and benefits of MBV to the educational organizations where individuals construct society. In this study, it is aimed to build an application plan of MBV for ideal educational organizations where students can reveal their potential, learn to unite around common values and have a fair management approach. In the application phases of the study, results showed some facts.

The first fact is that even the concept of MBV has not known well, students, academicians, and managers support this new management tool. In the study, we focused on the benefits of MBV, finding common values in the organization, finding factors that affecting a sense of belonging, and creating an application plan. However, we should give more effort into explaining the MBV to the students, academicians, and managers to raise awareness and increase existing support. Also, we should pay more attention to physically applying common values in the organization. The second fact is that the class of students and the title of academicians/managers have a reverse relation with the sense of belonging to the ITU. This result is supported by the research finding of Doğan (2016). This relationship called a degree factor in the study. It showed that when the expectations of the members not met, a sense of belonging is decreasing, which affects organizations negatively. Organizations should apply MBV for understanding and responding to the insufficiencies, requirements, and expectations of members. The third fact is that desired values are quality, innovation, globalism, and diligence, while insufficient values are equality and objectivity in ITU. Corresponding actions were listed for these values to be able to apply them. However, the study had done with 128 members of ITU, which should be increased. All of the members of the university should be participated in the survey to find more general, valid, and reliable results. The fourth fact is that managers have the most significant role in the application of MBV in organizations. In the study, we focused on the degree factor because of its effect on a sense of belonging. So, we did interviews with academicians from different degrees. However, when taking into consideration the authority and power of the managers, it could be considered to do interviews with managers to create more influence. Vurgun and Öztop (2011) supported this result in their studies and clarified that managers have the most significant responsibility and authority in understanding and applying the values.

The study shows essential and beneficial facts about MBV in engineering education. The interest, requests, and support of the members about MBV have been revealed. An application plan of MBV and its critical points constructed and identified. The importance and requirement of MBV had clarified. While technology is developing rapidly in the world, systems and processes are changing rapidly, and people's expectations are increasing rapidly, this is very important for engineering education to use MBV in order to be able to respond to all these conditions and to become one of the prominent ones instead of getting lost in this process.

REFERENCES

Akdemir, D. A. (2007). *Yönetici Değerlerinin, Eğitim Örgütleri Karar Sürecine Etkileri* (Unpublished master's thesis). Fırat Üniversitesi, Elazığ.

Altınkurt, Y., & Yılmaz, K. (2010). Değerlere Göre Yönetim ve Örgütsel Adalet İlişkisinin Ortaöğretim Okulu Öğretmenlerinin Algılarına Göre İncelenmesi. *Kuram ve Uygulamada Eğitim Yönetimi.*, *16*(4), 463–484.

Anonymous. (2002). Managing by values. *Strategic Dimension*, (22).

Beck, V. (2014). The effects of the implementation of value-based management. *International Journal of Economic Sciences and Applied Research.*, *7*(2), 153–165.

Bell, J., & Harrison, B. T. (2018). *Vision and Values in Managing Education: Successful Leadership Principles and Practice*. Abingdon, UK: Published by Routledge. doi:10.4324/9781351041508

Benedek, A., & Molnar, G. (2015, July). *E-teaching and Digitalization at BME*. Paper presented at the 19th International Conference on Engineering Education, Zagreb.

Blanchard, K. (1997). *Managing by values*. Executive Excellence.

Bukvić, V. (2016). Value-based management with some practical examples in Slovenian industries. *Advanced in Business-Related Scientific Research Journal, 7*(2).

Creswell, J. W. (2013). *Research Design (International Student Edition): Qualitative, Quantitative, and Mixed Methods Approaches* (4th ed.). Sage Publications, INC.

Demirtaş, Z. & Kılıç, M.F. (2015). Okul yöneticilerinin değerlerle yönetim kavramına ilişkin algıları. *Turkish Journal of Educational Studies*, (2).

Doğan, S. (2016). Model of values-based management process in school: A mixed design study. *Sciedu Press.*, *5*(1), 83–96.

Dolan, S. L., & Altman, Y. (2012). Managing by values: The leadership spirituality connection. *People and Strategy*, *35*(4), 20–26.

Dolan, S. L., & Garcia, S. (1999). *La Gestion par Valeurs: Una Nowelle Culture pour les Organisations*. Montreal: Editions Nouvelles.

Dolan, S. L., & Garcia, S. (2002). Managing by values cultural redesign for strategic organizational change at the dawn of the twenty-first century. *Journal of Management Development*, *21*(2), 101–117. doi:10.1108/02621710210417411

Dolan, S. L., & Raich, M. (2013). Coaching by values, entrepreneurship, and care A framework for reengineering an innovative and sustainable culture. *Kindai Management Review*, *1*, 2186–6961.

Dolan, S. L., & Richley, B. A. (2006). MBV: a new order philosophy for a new economic order. Handbook of Business Strategy.

Hair, J. F., Anderson, R. E., Tatham, R. L., & Black, W. C. (1995). *Multivariate Data Analysis with Readings* (4th ed.). Prentice-Hall Internatimal, Inc.

Harung, H. S., & Dahl, T. (1995). Increased productivity and quality through management by values: A case study of Manpower Scandinavia. *The TQM Magazine*, *7*(2), 13–22. doi:10.1108/09544789510081063

Haynes, F. (1998). *Ethical School*. London: Routledge Press.

Jaakson, K. (2010). Management by values: Are some values important than others? *Journal of Management Development*, *29*(9), 795–806. doi:10.1108/02621711011072504

Kerwin, S. & MacLein, J. & Bell-Laroche, D. (2014). The value of managing by values a Canadian sports organization case study. *Journal of Applied Sport Management*, *16*(4).

Krzakıewıcz, K. (2012). Management by values – a strategic dimension. *Management*, *16*(2), 7–15. doi:10.2478/v10286-012-0051-3

Krzakıewıcz, K. (2012). Management by values – a strategic dimension. *Management*, *16*(2), 7–15. doi:10.2478/v10286-012-0051-3

Peters, T. J., & Waterman, R. H. (1982). *In Search of Excellence: Lessons from America's Best-run Companies*. New York, NY: Harper & Row.

Peters, T. J., & Waterman, R. H. (1982). *In Search of Excellence: Lessons from America's Best-run Companies*. New York, NY: Harper & Row.

Polat, S. (2006). *Kurumsal Kültür Organizasyonlar Kurallar Kurumlar*. SPK.

Rokeach, M. (1983). *The nature of human values*. New York: The Free Press.

Schein, E. (1992). *Organizational Culture and Leadership*. New York, NY: Random House.

Šundov, J., & Dulčić, Z. (2017). Managing by values – Top managers impact on the process of creating new values for the company. *5th International OFEL Conference on Governance, Management, and Entrepreneurship The Paradoxes of Leadership and Governance in the Postmodern Society*.

Krzakıewıcz, K. (2012). Management by values – a strategic dimension. Management, 16(2).

Peters, T. J., & Waterman, R. H. (1982). *In Search of Excellence: Lessons from America's Best-run Companies*. New York, NY: Harper & Row.

Polat, S. (2006). *Kurumsal Kültür Organizasyonlar Kurallar Kurumlar*. SPK.

Rokeach, M. (1983). *The nature of human values*. New York: The Free Press.

Schein, E. (1992). *Organizational Culture and Leadership*. New York, NY: Random House.

Schuster, K., Grob, K., Vossen, R., Richert, A., & Jeschke, S. (2017). Preparing for Industry 4.0 – Collaborative Virtual Learning Environments in Engineering Education. In *Engineering Education 4.0*. Cham: Springer.

Šundov, J., & Dulčić, Z. (2017). Managing by values – Top managers impact on the process of creating new values for the company. *5th International OFEL Conference on Governance, Management, and Entrepreneurship The Paradoxes of Leadership and Governance in the Postmodern Society*.

Töremen, F., & Akdemir, D. A. (2007). Eğitim örgütlerinin kişiliğine uygun paradigma: Değerlerle yönetim. *E-Journal of New World Sciences Academy*, *2*(4), 1306–1311.

Vurgun, L., & Öztop, S. (2011). Yönetim ve örgüt kültüründe değerlerin önemi. *Süleyman Demirel Üniversitesi İktisadi ve İdari Bilimler Fakültesi Dergisi*, *16*(3), 217–230.

Walters, D. (2002). *Value and value chains in educations. Operations Strategy*. Red Globe Press.

ADDITIONAL READING

Dolan, S. L. (2012). *Managing by Values: The Leadership spirituality connection*. Global Future of Work Foundation – GFWF.

Dolan, S. L., Garcia, S., & Auerbach, A. (2003). Understanding and Managing Chaos in Organisations. *International Journal of Management*, *20*(1), 23.

Holmes, M. (1992). The Revival of School Administration: Alasdair MacIntyre in the Aftermath of the Common School. *Canadian Journal of Education*, *17*(4), 4. doi:10.2307/1495438

Koller, T. (1994). *What is value-based management? An excerpt from Valuation: Measuring and Managing the Value of Companies* (2nd ed.). The McKınsey Quarterly.

Milliman, J., Fergusan, J., Trickett, D., & Condemi, B. (1999). Spirit and community at Southwest Airlines: An investigation of spiritual values-based model. *Journal of Organizational Change Management*, *12*(3), 221–233. doi:10.1108/09534819910273928

Valle, R. C., Esteve, M., & Grau, M. (2013). Assessing public sector values trough the tri-axial model: Empirical evidence from Spain. *Cross Cultural Management, 20*(4), 528–543. doi:10.1108/CCM-03-2013-0045

Zhang, Y., Dolan, S., & Zhou, Y. (2009). Management by values: A theoretical proposal for strategic human resource management in China. *Chinese Management Studies, 3*(4), 272–294. doi:10.1108/17506140911007468

Zhen, L. (2012). Management by Values: A Case Study. *International Business and Management, 4*(2), 75–91.

KEY TERMS AND DEFINITIONS

Behavioral Sciences: A branch of science that examines and generalizes human behavior/action in different perspectives such as psychology, sociology, law, or management.

Commitment: Separating your time and interest by taking responsibility for your wishes and beliefs for the person, organization, or work you are affiliated with.

Educational Organizations: Institutions that form the foundations of the individual, provide connections between individuals, broaden the horizons of people, educate and develop the society, provide to gain perspective, and shape the foundations of thought.

Human Resource Management: Strategic approach aiming to bring out the highest potential of individuals and organizations by enabling people to work efficiently by guiding people in an organization more effectively.

Management Styles: The ways of carrying out activities such as planning, guiding, or organizing to ensure that the works and people in the organization achieve the success of the organization by achieving the objectives.

Organizational Culture: Refers to common values, beliefs, experiences, and actions which represent the organization uniquely.

Values: A set of principles that shape and control the behaviors of people. These are general concepts that reflect the opinions, emotions, priorities, and objectives of the majority to ensure the continuity of the organizations.

Values-Based Management: A new strategic leadership instrument that is a different way of sensing, processing, and applying information concerning behavioral sciences.

APPENDIX

A. Online Survey of Students – Quantitative Part

ITU Faculty of Management - Survey Study

With the rapid change after the twentieth century, the concept of management by values has existed due to the insufficiency of management by inspections and management by objectives. Management by values is a new strategic leadership tool that provides a collective spirit and value-oriented work by combining members in common values.

1. Choose your gender (Multiple choices)
   Female, Male
2. Choose your department (Multiple choices)
   Industrial engineering, Management engineering, Economics
3. Choose your grade (Multiple choices)
   1st grade, 2nd grade, 3rd grade, 4th grade
4. Choose your opinion about management by values (Multiple choices)
   I agree-MBV will support ITU, I don"t agree- MBV will not support ITU
5. Choose the level of belonging to ITU in ITU members. (1: Min, 5: Most) (Multiple choices)
   1, 2, 3, 4, 5
6. The values in the literature are listed below. Choose the values which are clearly shown in ITU. (You can select multiple values.) (Checkboxes).
   Quality, innovation, sensitivity, evolution, equality, reliability, flexibility, respect, multiculturalism, diligence, objectivity, national values, principles of Atatürk, cooperation, tolerance, globalism, other, (Demirtaş and Kılıç, 2015; Jaakson, 2010; Doğan, 2016; Harung and Dahl, 1995; Šundov and Dulčić, 2017).
7. Choose the four values that you want to represent ITU. (Checkboxes)
   Quality, innovation, sensitivity, evolution, equality, reliability, flexibility, respect, multiculturalism, diligence, objectivity, national values, principles of Atatürk, cooperation, tolerance, globalism, other, (Demirtaş and Kılıç, 2015; Jaakson, 2010; Doğan, 2016; Harung and Dahl, 1995; Šundov and Dulčić, 2017).
8. Share your suggestions and thoughts. (Open-ended).

B. Online Survey of Academicians/ Managers – Quantitative Part

ITU Faculty of Management - Survey Study

With the rapid change after the 20th century, the concept of MBV has existed due to the insufficiency of management by inspections and management by objectives. Management by values is a new strategic leadership tool that provides a collective spirit and value-oriented work by combining members in common values.

1. Choose your gender (Multiple choices)
   Female, Male
2. Choose your task (Multiple choices)
   Academician, Administrative officer academician (Manager)
3. Choose your department (Multiple choices)
   Industrial engineering, Management engineering, Economics
4. Select the year range that you work as an academician. (Multiple choices)
   1-5 years, 6-10 years, 11-15 years, 16 years and above
5. Select the year range that you are the administrator. (This question is only for academicians who have the duty of the manager.) (Multiple choices)
   1-3 years, 4-6 years, 7-9 years, ten years and above
6. Choose the title which you have. (Multiple choices)
   Research assistant, Lecturer doctor, Doctor lecturer, Associate professor, Professor doctor
7. Choose your opinion about management by values (Multiple choices)
   I agree – MBV will support ITU, I don"t agree - MBV will not support ITU
8. Choose the level of belonging to ITU in ITU members. (1: Min, 5: Most) (Multiple choices)
   1, 2, 3, 4, 5
9. The values in the literature are listed below. Choose the values which are clearly shown in ITU. (You can select multiple values.) (Checkboxes).
   Quality, innovation, sensitivity, evolution, equality, reliability, flexibility, respect, multiculturalism, diligence, objectivity, national values, principles of Atatürk, cooperation, tolerance, globalism, other, (Demirtaş and Kılıç, 2015; Jaakson, 2010; Doğan, 2016; Harung and Dahl, 1995; Šundov and Dulčić, 2017).
10. Choose the four values that you want to represent ITU. (Checkboxes)
    Quality, innovation, sensitivity, evolution, equality, reliability, flexibility, respect, multiculturalism, diligence, objectivity, national values, principles of Atatürk, cooperation, tolerance, globalism, other, (Demirtaş and Kılıç,

2015; Jaakson, 2010; Doğan, 2016; Harung and Dahl, 1995; Šundov and Dulčić, 2017).

11. Share your suggestions and thoughts. (Open-ended)

C. Interview of Academicians – Qualitative Part

ITU Faculty of Management - Interview Study

1. "The MBI and MBO are replaced by MBV in all over the world due to a rapid and uncertain environment. Organizations are aiming to increase their success and loyalty by constructing common values and taking their decisions for these common values. Based on the survey result, more than 90% of academicians and %80 of students think MBV is beneficial for ITU." What does management by values mean to you?

2. "The results of surveys show that current values are existing less than desired rates. Some of the participants specified that values are not using anymore in management of ITU." Do you think any attention is paid to values in the management of ITU? Do you think there is a difference between the past and present situation of management in ITU? Why? What should be done?

3. "The results of surveys show that most desired values in the ITU are quality, innovativeness, globalism, and diligence. Insufficient values are equality and objectivity. " Do you think the desired values are logical and applicable? Why? Can you suggest application examples for desired values? What are the reasons for the insufficient values? What should be done? Do you want to add any value to common values?

4. "Most important persons in the application process of MBV are managers of the organizations. Common values can be applied, disseminated, and clarified successfully and permanently with the collaboration of managers. Based on surveys, the common factor in students and academicians, which affecting the belonging rate, is the degree. When the class of students is increasing, and the degree of academicians is increasing, belonging rate to ITU is decreasing." What are the reasons for these results? How can we apply management by values in ITU? Can you make a recommendation to the application options?

D. Results of Regression Analyses of Students' Survey – Quantitative Part

Figure 5. Variables entered and removed from students

Variables Entered/Removed[a]

Model	Variables Entered	Variables Removed	Method
1	Class, Eco, Gender, ManEng[b]	.	Enter
2	.	ManEng	Backward (criterion: Probability of F-to-remove >= ,100).
3	.	Eco	Backward (criterion: Probability of F-to-remove >= ,100).
4	.	Gender	Backward (criterion: Probability of F-to-remove >= ,100).

a. Dependent Variable: Belonging

b. Tolerance = ,000 limit reached.

Figure 6. Model summary

Model Summary

Model	R	R Square	Adjusted R Square	Std. Error of the Estimate
1	,708[a]	,501	,479	,71788
2	,708[b]	,501	,485	,71398
3	,706[c]	,498	,487	,71203
4	,703[d]	,494	,489	,71113

a. Predictors: (Constant), Class, Eco, Gender, ManEng

b. Predictors: (Constant), Class, Eco, Gender

c. Predictors: (Constant), Class, Gender

d. Predictors: (Constant), Class

Figure 7. ANOVA of students

ANOVA[a]

Model		Sum of Squares	df	Mean Square	F	Sig.
1	Regression	47,061	4	11,765	22,829	,000[b]
	Residual	46,898	91	,515		
	Total	93,958	95			
2	Regression	47,059	3	15,686	30,772	,000[c]
	Residual	46,899	92	,510		
	Total	93,958	95			
3	Regression	46,808	2	23,404	46,162	,000[d]
	Residual	47,150	93	,507		
	Total	93,958	95			
4	Regression	46,423	1	46,423	91,799	,000[e]
	Residual	47,536	94	,506		
	Total	93,958	95			

Figure 8. Coefficients of students

Coefficients[a]

Model		Unstandardized Coefficients B	Unstandardized Coefficients Std. Error	Standardized Coefficients Beta	t	Sig.
1	(Constant)	4,611	,241		19,145	,000
	Gender	,132	,151	,067	,875	,384
	ManEng	,009	,187	,004	,048	,962
	Eco	-,108	,178	-,050	-,604	,547
	Class	-,626	,069	-,717	-9,112	,000
2	(Constant)	4,617	,197		23,391	,000
	Gender	,131	,149	,066	,880	,381
	Eco	-,112	,159	-,052	-,702	,484
	Class	-,627	,066	-,718	-9,531	,000
3	(Constant)	4,581	,190		24,120	,000
	Gender	,130	,149	,065	,872	,386
	Class	-,625	,066	-,717	-9,541	,000
4	(Constant)	4,615	,186		24,874	,000
	Class	-,614	,064	-,703	-9,581	,000

a. Dependent Variable: Belonging

E. Result Tables of Regression Analyses of Academicians' /Managers' Survey – Quantitative Part

Figure 9. Variables entered and removed from academicians/managers

Variables Entered/Removed[a]

Model	Variables Entered	Variables Removed	Method
1	Degree, ManEng, Gender, ManAca, IndEnd, ManYears, AcaYears[b]	.	Enter
2	.	ManYears	Backward (criterion: Probability of F-to-remove >= ,100).
3	.	Gender	Backward (criterion: Probability of F-to-remove >= ,100).
4	.	IndEnd	Backward (criterion: Probability of F-to-remove >= ,100).
5	.	AcaYears	Backward (criterion: Probability of F-to-remove >= ,100).

a. Dependent Variable: Belonging

b. Tolerance = ,000 limit reached.

Figure 10. Model summary of academicians/managers

Excluded Variables[a]

Model		Beta In	t	Sig.	Partial Correlation	Collinearity Statistics Tolerance
1	Eco	.[b]	.	.	.	,000
2	Eco	.[c]	.	.	.	,000
	ManYears	-,068[c]	-,295	,771	-,060	,488
3	Eco	.[d]	.	.	.	,000
	ManYears	-,058[d]	-,258	,799	-,051	,493
	Gender	,068[d]	,385	,704	,077	,816
4	Eco	-,133[e]	-,808	,426	-,157	,907
	ManYears	-,099[e]	-,456	,652	-,089	,528
	Gender	,017[e]	,100	,921	,020	,917
	IndEnd	,132[e]	,808	,426	,157	,910
5	Eco	-,106[f]	-,663	,513	-,127	,943
	ManYears	-,123[f]	-,587	,562	-,112	,549
	Gender	,039[f]	,242	,811	,047	,959
	IndEnd	,106[f]	,663	,513	,127	,946
	AcaYears	-,181[f]	-,691	,495	-,132	,350

a. Dependent Variable: Belonging

b. Predictors in the Model: (Constant), Degree, ManEng, Gender, ManAca, IndEnd, ManYears, AcaYears

c. Predictors in the Model: (Constant), Degree, ManEng, Gender, ManAca, IndEnd, AcaYears

d. Predictors in the Model: (Constant), Degree, ManEng, ManAca, IndEnd, AcaYears

e. Predictors in the Model: (Constant), Degree, ManEng, ManAca, AcaYears

f. Predictors in the Model: (Constant), Degree, ManEng, ManAca

Figure 11. ANOVA of academicians/managers

ANOVA[a]

Model		Sum of Squares	df	Mean Square	F	Sig.
1	Regression	9,679	7	1,383	2,033	,092[b]
	Residual	16,321	24	,680		
	Total	26,000	31			
2	Regression	9,620	6	1,603	2,447	,053[c]
	Residual	16,380	25	,655		
	Total	26,000	31			
3	Regression	9,523	5	1,905	3,006	,029[d]
	Residual	16,477	26	,634		
	Total	26,000	31			
4	Regression	9,109	4	2,277	3,640	,017[e]
	Residual	16,891	27	,626		
	Total	26,000	31			
5	Regression	8,810	3	2,937	4,784	,008[f]
	Residual	17,190	28	,614		
	Total	26,000	31			

a. Dependent Variable: Belonging

b. Predictors: (Constant), Degree, ManEng, Gender, ManAca, IndEnd, ManYears, AcaYears

c. Predictors: (Constant), Degree, ManEng, Gender, ManAca, IndEnd, AcaYears

d. Predictors: (Constant), Degree, ManEng, ManAca, IndEnd, AcaYears

e. Predictors: (Constant), Degree, ManEng, ManAca, AcaYears

f. Predictors: (Constant), Degree, ManEng, ManAca

Figure 12. Coefficients of academicians/managers

Coefficients

Model		Unstandardized Coefficients B	Std. Error	Standardized Coefficients Beta	t	Sig.
1	(Constant)	3,820	,561		6,813	,000
	Gender	,136	,335	,073	,406	,688
	IndEnd	,259	,332	,143	,780	,443
	ManEng	1,805	1,009	,348	1,788	,086
	ManAca	,936	,502	,377	1,865	,074
	AcaYears	-,163	,257	-,185	-,635	,531
	ManYears	-,057	,192	-,068	-,295	,771
	Degree	-,154	,159	-,275	-,969	,342
2	(Constant)	3,860	,533		7,241	,000
	Gender	,126	,327	,068	,385	,704
	IndEnd	,279	,318	,155	,877	,389
	ManEng	1,796	,990	,347	1,813	,082
	ManAca	,862	,427	,347	2,020	,054
	AcaYears	-,183	,244	-,207	-,750	,460
	Degree	-,158	,155	-,283	-1,019	,318
3	(Constant)	3,941	,482		8,176	,000
	IndEnd	,239	,296	,132	,808	,426
	ManEng	1,697	,940	,328	1,804	,083
	ManAca	,851	,419	,343	2,032	,053
	AcaYears	-,197	,237	-,224	-,831	,413
	Degree	-,147	,150	-,264	-,981	,335
4	(Constant)	4,030	,466		8,645	,000
	ManEng	1,658	,933	,320	1,776	,087
	ManAca	,852	,416	,343	2,049	,050
	AcaYears	-,160	,231	-,181	-,691	,495
	Degree	-,176	,145	-,316	-1,219	,234
5	(Constant)	3,777	,285		13,256	,000
	ManEng	1,982	,799	,383	2,480	,019
	ManAca	,885	,409	,357	2,163	,039
	Degree	-,253	,092	-,453	-2,756	,010

a. Dependent Variable: Belonging

Section 2

Chapter 6
Use of Collaborative Technologies in Engineering Education

Hasan Çakır
https://orcid.org/0000-0002-4499-9712
Gazi University, Turkey

Erhan Ünal
Afyon Kocatepe University, Turkey

ABSTRACT

The purpose of this chapter is to explain the collaborative problem-solving approach and collaborative technologies that help engineering students to establish and improve collaboration in their coursework. To this end, the theoretical background of collaboration in education and the importance of the learning environments are discussed. Possible effects of a constructivist learning environment on engineering students' educational output are explained. Following that, the collaborative problem-solving approach and collaborative technologies are presented. Then, the collaborative problem-solving method framework and how collaborative technologies can be used with this method in the learning environment of engineering education are explained in detail. Finally, recommendations about future work are presented.

INTRODUCTION

In every context of education, students' meaningful engagement with sound academic activities improves their educational outcomes such as persistence with the school,

DOI: 10.4018/978-1-7998-2562-3.ch006

academic achievement, completing the enrolled program, and personal development (Carini, Kuh, & Klein, 2006). Engineering education is not an exception (Ohland, Sheppard, Lichtenstein, Eris, Chachra & Layton, 2008). The level of student engagement and student success with the engineering education programs also affect schools' reputation and investment plans, therefore it is important to increase the students' involvement with the school and their success with staying in school and completing the degree (Chen, Lattuca, & Hamilton, 2008). Since engineering education is a tertiary education level, it is important to recognize Tinto's (1987) and Astin's (1993) works on student involvement and engagement with academic activities in higher education.

Tinto (1987) developed a model for factors affecting students' decision to drop out of college, which is called the institutional departure model. The purpose of the model is to explain how factors affect and interact with students' decisions on whether to complete or depart from higher education. As shown in Figure 1, five major factors impact this decision, one of which is institutional experience and others are personal traits. Institutional experience is the only factor that allows faculty, administrator, and policymakers to implement in learning environments for improvement in students' experiences with higher education. Peer interactions, faculty/staff interactions, and extracurricular activities help students to have positive experiences with higher education.

Figure 1. Tinto's Model of Institutional Departure (Tinto, 1993)
(Source: Tinto, V. (1993). Leaving college: Rethinking the causes and cures of student attrition. 2nd edition. Chicago: The University of Chicago Press, p.114)

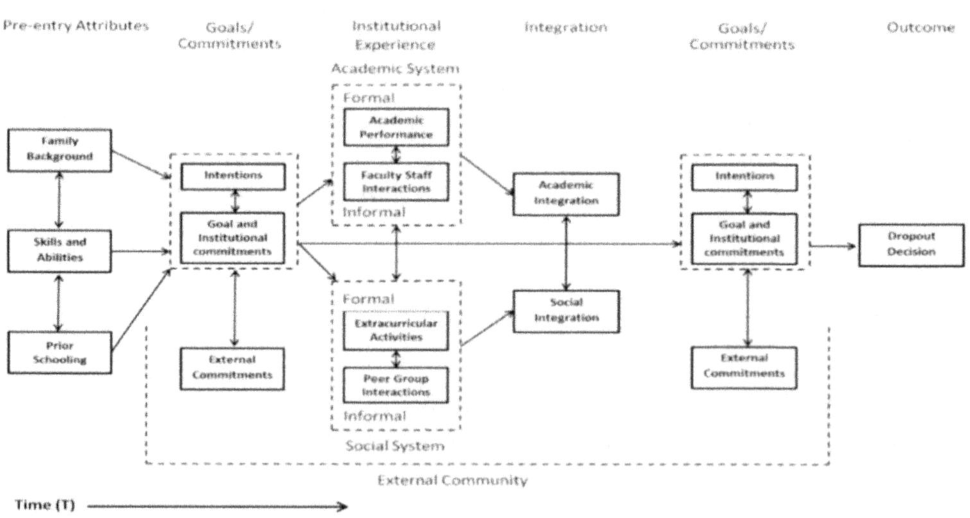

Astin (1993) developed a theory of student involvement with undergraduate education. He explains that student involvement with higher education requires continuous psychosocial and physical energy, both quantity and quality of the higher education activities can be counted towards involvement, student personal development and student academic achievement positively correlate with student involvement. Astin's theory points out that during their education, higher education students should be provided with as many social, physical, and instructional opportunities for involvement as possible. By doing so, higher education experience contributes to their personal development and success in their life.

Theories of student retention and student involvement contain implications for designing instruction and learning environments at higher education. Considering thirty years of research on student retention and student involvement in higher education, Chickering and Gamson (1987) developed seven principles for good practice in undergraduate education. These principles were instructional recommendations to improve student involvement and retention. Since their publication, they were the inspiration for practices in higher education, online education and student engagement in many research studies. What are these seven principles? According to Chickering and Gamson (1987) good practice in undergraduate education;

1. Encourages contact between students and faculty,
2. Develops reciprocity and cooperation among students,
3. Encourages active learning,
4. Gives prompt feedback,
5. Emphasizes time on task,
6. Communicates high expectations,
7. Respects diverse talents and ways of learning,

As one can see, all these principles encourage interactions among students and between faculty and students, value active learning activities, and multiple methods of instruction. The implementation of these principles with sound instructional strategies improves students' experiences with higher education and their success. In order to implement all these principles in learning environments at higher education levels, the social constructivist approach to learning environment design can be considered as one of the strategies for instructional design. The social constructivist approach to instruction is based on the constructivist view of learning and encourages intense collaboration among students within the different stages.

Theoretical Background

Collaboration is viewed as making decisions based on a group of people's common understanding of the problem at hand with shared goals and values. Rather than combining pieces of information or different individual's work to make a decision, it requires group members to have dialogue/discussions and reach consensus related to the problems that need to be solved. Therefore principles of the constructivist and social constructivist view of learning provide a basis for the theoretical background.

Constructivist explanation of learning argues that the human mind constantly seeks for meaning and is not an empty box that should be filled with adults' knowledge (Duffy & Cunningham, 1996). Using sensor organs, the human mind perceives the outside world and creates meaning and explanations for the encountered reality in conjunction with existing knowledge in the mind. By doing so, *a web of knowledge is created in an individual's mind*, which is this particular individual's truth. If there is a conflict between perceived reality in the outside world and the truth residing in the web of knowledge in an individual's mind, the perceived reality does not fit in the existing web of knowledge. This causes the individual to be perplexed and s/he starts to seek more information, meaning, and explanation to accommodate the newly encountered situation to the existing web of knowledge by enhancing pre-existing knowledge in his/her mind (Duffy & Cunningham, 1996).

Continuing on individual cognition of the constructivist view of learning, in another perspective of the constructivist approach to learning, Vygotsky (1980) argues that individuals do not live in isolation rather they live in groups or societies. They interact with one another using communication tools, for example, their language to get their messages across. In the *social view of constructivism*, individuals create meaning of reality by communicating and interacting with each other meaning that they carry on discussions to *reach a common understanding of the perceived reality*. The consensus they reach about perceived reality by discussing becomes their shared truth, therefore, the web of knowledge lies in the society rather than an individual's mind (Duffy & Cunningham, 1996; Jonassen, 1999). For example traditions, etiquette rules, laws, or regulations are all shared consensus and knowledge of a group of people living in the same environment and working towards similar goals. They were created by years of experience and discussions among the members of the society where they were accepted as shared knowledge. They can be acceptable by the society in which they were created but may not be acceptable by other societies. Therefore, the social constructivist view of learning assumes that *the truth generated by reaching consensus is relative to the society* in which it is created.

Considering the above very succinct explanation of the social constructivist view of learning three implications can be derived for learning environments;

Figure 2. Elements of Constructivist Learning Environment (Jonassen, 1999)
Source: Designing constructivist learning environments. In Reigeluth, C. M. (Ed.), Instructional-Design Theories and Models, Vol. II (pp. 215-239). New Jersey: Lawrence Erlbaum Associates, p. 218)

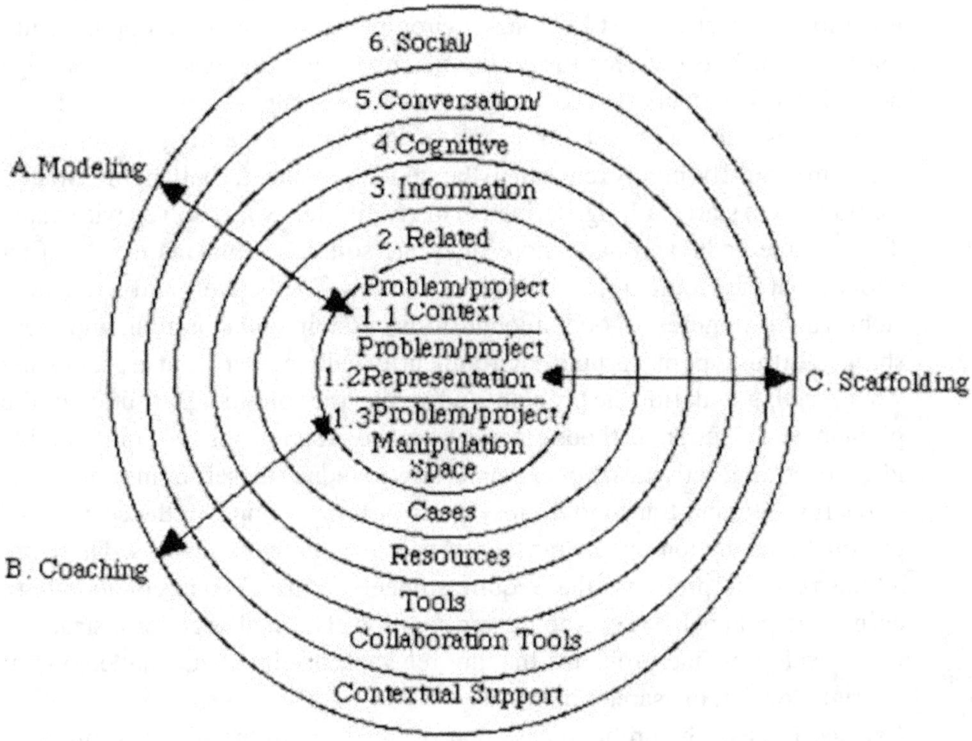

1. Students should encounter an unstructured perplexing problem situation that is perceived differently by all of them and there is no one solution or one true answer. It serves the purpose of starting to seek new information and ignite discussions.
2. Students need to have communication and collaboration tools to discuss the problem at hand and to generate/share ideas related to the definition of the problem and possible solution methods.
3. Students need to have communication and collaboration tools to receive feedback from experts about their understanding of the problem and their solution methods.

By considering all principles of the social constructivist view of learning, Jonassen (1999) offered a framework to explain elements that should be developed in a constructivist learning environment for students. Figure 2 conceptually depicts these elements and their relations to each other.

As shown in Figure 2, a constructivist learning environment should have;

1. **Unstructured Problem/Project Situation**: An unstructured problem situation is central to constructivist learning environments. It provides an opportunity to create conflicts between the reality perceived and pre-existing knowledge of students. Unstructured problems do not have straight forward definitions or solutions and it serves the purpose of igniting the learning process. The problem should be highly relevant to the students' context, challenging enough for students to start working, defined in many different ways, solved with many different methods, and have more than one solution. Students need to first define what the problem is and how it can be solved effectively and efficiently. Achieving it requires a good amount of discussion, collaboration, and idea-sharing. In this type of learning environment, usually, students put in groups and work together to define the problem and to plan the solution. Definition of the problem and solution methods emerge from the group members' collaborative idea-generating activities. For example, how to reduce diesel engine emissions or energy-efficient transportation systems or how to improve the security of personal information that is transferred on the computer networks. These are all unstructured problems that require students to discuss and collaborate to define and plan before developing any solution. The challenge for instructors is to develop or find problems that are relevant, original, and challenging at the right level at the same time.
2. **Related Cases**: The problem case that is presented to the students might be previously studied by others or perhaps there were developed solutions in similar or related cases. For example, the problem of developing low emission vehicles for mechanical engineering students is not something they encounter for the first time. There might be others who worked on the same problem and developed solutions that worked best for them. By comparing and contrasting the previously defined and solved problems with the problem at their hand, students define the problem and develop their solutions that are highly likely to solve their problem.
3. **Information Resources**: In order to define and solve an unstructured problem, students need to expand their existing knowledge by acquiring knowledge from sources. Books, journal articles, instructional materials, databases, case studies, and all other information resources help students to acquire new knowledge to solve the problem. The internet has become a great help for instructors to provide information sources for students.
4. **Cognitive Tools:** In order to develop a solution for the given unstructured problem, students should have the opportunity to try out their ideas or hypotheses about what could be the possible solutions. For example, an engine

simulator or a computer network simulation software allows students to try their solutions for given problems. Cognitive tools help students to enhance their cognitive abilities to calculate, to generate rationality, and to test the solutions. Calculation tools, simulation tools, or any tools that help students to develop their solutions are considered as cognitive tools.

5. **Conversation and Collaboration tools:** As described at the beginning of this section, the social construction of knowledge requires a lot of discussion, idea sharing, and reaching a consensus. Therefore, the students need to be provided with platforms and/or tools that allow them to exchange and share ideas, keep the records of discussions, and collaborate on generating documents or products during the problem-solving process. With the development of dynamic web technologies, the internet and World Wide Web have immense potential to provide online communication and collaboration tools for students and instructors. Online chat tools, discussion platforms, collaborative document creation tools, project planning, and collaborative software development tools are just a few examples of platforms that allow students to work collaboratively on problems.
6. **Social/Contextual Support:** Students that are put in groups belong to a larger community and this community consists of other students, instructors, and experts. While working in groups on the given unstructured problem, the students may reach a common understanding of the problem and may develop a solution that they think is the answer to the problem. However, they need to validate their knowledge about the problem and the solution by asking opinions of the greater community. Fellow students in the classroom, instructors or experts in the community may review their work and give feedback. Implementing this activity also requires the use of communication and collaboration tools.
7. **A Mentor with Modeling, Coaching, and Scaffolding Roles:** Constructivism argues that each individual constructs his/her knowledge, however, it does not mean that no or little guidance is provided. Unlike roles and responsibilities in a traditional learning environment, instructors' roles in constructivist learning environments can be grouped into three major categories; Modeling, coaching, and scaffolding. In modeling role, instructor and student work as a master-apprentice team, instructor and students work on the same problem together and students learn the process of problem-solving by mimicking or watching as they work on the problem with the master. Coaching refers to guiding students to a desired solution without giving the solution to them. During the problem definition and solution generation process, students may wander around and digress from the main aim of the learning process. By asking guiding questions such as "have you thought in this angle", "do you think there is a better solution to this", "have you check this person's previous work", instructors help students

to stay focused on the solution. Lastly, scaffolding is a means to help students to be able to start with tackling the problem. Not all students in a class may have pre-knowledge and skills to work on the problem, therefore instructors should find tactics to cognitively support them to a point where they can start working on the given problem.

Constructivist learning environments give instructors an opportunity to implement a collaborative problem-solving process in their classroom. The idea of having students work in groups and collaborate with peers and the larger community is a sound instructional practice in terms of academic engagement however it requires support from technology to provide a platform for collaboration, documentation, and interaction.

In order to implement constructivist learning environments in classroom settings, an instructional framework called the collaborative problem-solving process could be utilized. The collaborative problem-solving process is explained below in detail. It provides guiding principles for how to implement problem-based learning method with an emphasis on collaboration.

Engaging students with academically meaningful activities and contributing to their personal development and success is one of the methods to increase students' learning outcomes in their higher education experience (Kuh, Cruce, Shoup, Kinzie, & Gonyea, 2008). Engineering education students are not an exception. Additionally, due to the nature of engineering education, its curriculum contains hands-on experience, project-based learning, and design activities. The collaborative problem-solving process helps instructors to address all these aspects of the curriculum while students employ online collaboration tools and contribute to students' development in their education (Laird & Kuh, 2005).

Implementing constructivist learning environments and evaluating their impact on student outcomes has a long research tradition in problem-based learning research. Using problem-based learning as a main method of instruction has an impact on two major areas of student education outcomes; improving student engagement (Rotgans & Schmidt, 2011) and improving 21st century skills/competencies (Norman & Schmidt, 1992).

Once the students work in problem-solving or project design activities, their engagement with academically meaningful activities increases in comparison to traditional methods of instruction such as lecturing. They search for new information in information sources, write reports, collaborate with their group mates, and seek feedback from their instructors and peers. All these activities are called behavioral engagement of students with their course work. They also develop methods and skills to complete their project work effectively and efficiently which are called metacognitive skills or cognitive engagement. Problem-based learning or project-based learning gives students opportunities to work on a design problem and enhance their engagement with their academic program, which addresses all aspects of Chickering and Gamson's good practices for undergraduate education. As indicated at the beginning of chapter, student academic engagement is positively correlated with student success in higher education (Carini et al., 2006) hence implementing problem or project-based learning in higher education helps students to improve their success in higher education (Dolmans, Loyens, Marcq, & Gijbels, 2016; Korucu & Çakır, 2018; Topalli & Cagiltay, 2018; Yew & Goh, 2016)

The second major contribution of problem based or project-based learning is to help students to improve their 21st century learner competencies. The concept of 21st century learner skills has gained attention since much of teaching and learning activities are focused on helping learners to develop skills for metacognition, critical thinking, working together, and use of technology to do tasks (Chen, Wang, Kirschner, & Tsai, 2018; Chis, Moldovan, Murphy, Pathak, & Muntean, 2018; Koszalka, Song & Grabowski, 2001; Ravitz, 2009; Sendag & Odabasi, 2009). These skills can be gained and internalized only by practicing them over the period of undergraduate education. Implementing a problem-based or project-based learning method in engineering courses can help students to develop critical thinking skills to solve problems. It also helps them to develop skills for working together in groups and collaborate to achieve a learning or job related task. Finally, they use technological tools to complete the tasks effectively and efficiently in the problem solving process. Especially, they intensely use cognitive tools that are in the form of computer software and online collaboration tools, which improves students' technology use in the job place and integration of technology into work skills.

Considering the nature of engineering education and principles of academic engagement, implementing the collaborative problem solving process as instructional approach helps both instructors and students to achieve the objectives of the engineering curriculum, to improve interactions amongst instructors and students, to improve personal development of students, and to increase the overall success of the students and academic programs. The purpose of this chapter is to explain instructional methods and technologies that help engineering students to establish and improve collaboration in their coursework.

Collaborative Problem Solving

Collaborative problem solving is based on social constructivism and a more comprehensive approach than problem based learning and cooperative learning. This approach was evolved from problem solving process, collaborative learning theory, problem based and cooperative instructional theories. Collaborative problem solving is an approach that students are not only engaging in activities to solve a problem or create a project but also developing problem solving, critical thinking and team building skills (Nelson, 1999). In this regard, students can learn the knowledge and skills related to engineering and 21st century skills.

Collaborative problem solving comprised of collaboration and problem solving constructs. Collaboration is essential during problem solving. Given problems can be difficult to be solved individually. They can be solved by the collective work of group members. In order to achieve this, creating a heterogeneous group is essential. Group members have different experiences, pre-knowledge, and skills. Therefore all group members share resources, knowledge and make a contribution to problem solving (Fiore et. al, 2017).

Collaborative problem solving approach has some pedagogical values. This approach aims to enhance the collaborative processes of students. Learning environments should be student centered which supports students to be active participants in their own learning process. Furthermore, students should have the responsibility for their learning process. This environment improves students' content knowledge providing multiple perspectives and support the development of 21st skills. As a final value, this approach emphasizes social context and collaboration between students and instructors (Nelson, 1999). Thus, this approach is theoretically based on social constructivism.

In order to implement this approach in the learning environment of engineering, there are several issues should be taken into consideration. First of all, the collaborative problem solving approach is more suitable for heuristic tasks or tasks which have straight forward definitions or solutions. Thus students can improve conceptual understandings of the content and develop thinking skills, learning strategies and metacognitive skills. Second, the learning environment should support collaboration. In addition, necessary resources, time, and space should be provided. Therefore, students can engage in problem solving with a supportive climate of this collaborative work. Third, students should have ownership of the learning process. Finally, instructors should encourage students and facilitate students learning. Therefore instructors can be seen as a facilitator than a lecturer. For example, instructors prepare questions to focus students on the course content, organize and teach knowledge or skills that students need to complete the collaborative work (Nelson, 1998).

Collaborative Technologies

The advent of web technologies created new ways of learning and teaching. In 2004, the term "Web 2.0" was introduced by Tim O'Reilly. Web 2.0 technologies can be described as a new generation of web technologies due to the collective changes during the web is used (O'Reilly, 2005). According to Butler (2012), web 2.0 technologies can be defined as "a wide array of web-based applications which allow users to collaboratively build content and communicate with others across the world." (p.139). In the web 1.0 era, before the development of web 2.0 technologies, there were limited opportunities in terms of communication and interaction between authors on the web and users of the web. Thus, users were seen as consumers of the information on the web. Participative architecture is an important point for the development of web 2.0 technologies. In this regard, the role of individuals changed from consumers to active contributors. Thanks to web 2.0 technologies, individuals can add, share and edit different types of content such as text, image, voice, and graphics (Albion, 2008). In other words, web 2.0 technologies encourage users to participate in various activities such as creating, sharing and distributing the knowledge, communicating and collaborating with other users. As a result, web 2.0 technologies can be seen as a new philosophy of interaction on the web due to the opportunities they provided. Therefore, collaborative technologies term can be used for web 2.0 technologies.

Web 2.0 technologies are being used for different purposes. Although there are different web applications that can be used by individuals, an agreement on the classification of these technologies is limited. In general, web 2.0 technologies include social networks (Facebook, Edmodo), media sharing (Youtube, Flickr), social bookmarking tools (Delicious and CiteULike), collaborative authoring and knowledge construction tools (Wikipedia), content aggregation and organization tools (RSS, tagging tools), blogging and podcasting (Greenhow, Robelia, & Hughes, 2009). In another study, Bower (2016) examined over 200 web 2.0 technologies for teaching and learning. Then the researcher grouped them according to their characteristics. Finally, 14 clusters occurred for the typology of web 2.0 technologies. It can be said that the content type that web 2.0 technologies support and degree of synchronicity they provide determine the purpose of these technologies (Bower, Hedberg, & Kuswara, 2010). Therefore, some of the collaborative technologies are selected from previous classification studies mentioned above in order to use in engineering education for collaborative activities and are explained after using the related collaborative technology below.

Social Network

Social networks provide users to create online communities doing the following activities such as creating a personal profile, sharing any content (text, picture, video), participating in interest groups, making a comment to the postings. Facebook is a well-known social networking site. Edmodo is another social network site that can be seen as a specialized social network site for education. Thus instructors can create classes on Edmodo and get a class code. Then students who know the class code can access the class. Instructors can publish postings, share course resources (document, educational video, e-book and etc.), give homework, create quizzes, and conduct discussions via Edmodo. Twitter is another social network site that can be seen as a microblogging site. Youtube is another social network site for sharing videos. Youtube can be used in courses to support learning.

Blogging tools

Blogs are websites that can be published on the World Wide Web, where people can publish with different types of content, such as text, image, video, to share with other people in any subject. These contents name as post and the posts are organized in chronological order. In order to create a blog, users don't have to know any programming knowledge. Like website creation tools, blogs offer an editor to design the blog. Different services such as blogger.com, wordpress.com, tumblr.com, edublogs.org offer blog creation freely.

Wiki

Wikis enable users to create and edit websites in a collaborative manner. This website named as wiki. Users can create, edit and organize wiki pages on the internet. The advantage of the wiki is that users can edit wiki pages, add a hyperlink and then publish them. Therefore instructors can create wiki pages for students to work collaboratively. Different services as mediawiki.org, pbworks.com, wikia.com can be used to create a wiki page freely.

Mindmapping Tools

These tools are useful for creating a mind map for the course or project as the mindmap is a graphical representation of ideas and concepts. Therefore it can be used to make easier to learn the new concept and ideas. In order to create a meaningful mindmap, these tools can be taken into consideration. For example, Bubbl.us is one of the mindmapping tools. It allows users to create a text-based map with the ability to save the map as image format and colorful design. Mind42 is another free tool for creating a mind map. In addition, collaboration during creating mindmaps and adding images to the mindmaps are the advanced features of Mind42.

Infographic Tools

These tools help users to visualize the data with different styles as charts, graphs, and tables. Therefore different data types as text and numeric can be visualized with infographics. For example, Infogram is a tool to create powerful infographics for learning as different images and icons can be added and different data can be imported to the infographics. In addition, infographics can be exported with different formats. Venngage is another practical tool for creating infographics. Powerful infographics can be created and edited with this tool. As a result, interactive infographics as charts, graphs or reports can be used as a presentation or production of collaborative works or projects.

Document Creation Tools

These tools enable users to create and edit a document. Like an office document program on the computer, these tools provide similar features online. Users can edit the document synchronously. For example, Google Docs enable collaborative authoring of documents. Therefore students can create online documents or reports collaboratively in the courses. Besides, these documents can be stored on the other Google Application, Google Drive. Word Online is another similar tool. It is similar to the Google Docs. In addition, the created documents can be stored on the other Microsoft Application, OneDrive.

File Storing and Sharing Tools

Online file storing and sharing tools help users to store any folders or files on the web. In addition, users can share these folders or files with other users. Therefore other users can access them with the permission of the owner. These systems are useful for students and teachers during collaborative learning. For example, students can create a group folder and access the folder or file anywhere and anytime. Then each student can make a contribution to the group work. Google Drive, OneDrive and Dropbox are used for file storing and sharing. Furthermore, these systems incorporate online office programs. Thus online office tools can be executed in an integrated way on Google Drive, OneDrive and Dropbox.

Conferencing Tools

Online communication is easy and useful with web conferencing tools. These tools provide users text based, voice based or video based communication with other users. Therefore users can carry out different conferences on the internet. Online meetings can be used in the learning process and students can learn together. For example, Google Hangouts offers online meetings up to 10 users. Thus it can be seen as a supporting tool for collaborative works in the courses. Users can share pictures, text, the screen of the computer during the conference. In addition, other

Google Applications such as Google Docs, Google Slides can be used to complete a collaborative work.

Presentation Tools

Different styles, animations, images are necessary elements of a better presentation. In this regard, effective presentations can be created with the help of presentation tools. For example, Prezi is an online presentation tool. It is easy to create animated presentations via this tool. Different animated effects can be applied to the presentation. The final presentation can be shared with different users. Google Slides is an online presentation tool. It has similar properties as Prezi. But collaboration is an additional property of this tool. Users can create and edit a presentation collaboratively.

Website Creation Tools

These tools enable users to create free websites with having no web programming skills. Users can design a web page freely. For example, Google Sites is used for creating websites. To design a web page on Google sites, users should have a Google account. Then users can easily design the site such as determining the name of the site, adding text, image and multimedia files and publishing the site. All files related to the site are stored in Google Drive. Moreover, users can collaborate during designing the website. Weebly for Education is another easy website creating a tool for educational purposes. This tool differs from other tools offering educational features. Instructors can design educational websites easily with drag drop feature of the editor. This tool offers a secure environment for students. Instructors can adjust a password for the website. Therefore, only students can access the website.

Benefits of Collaborative Technologies

Collaborative technologies are web applications that allow users to create, analyze, exchange, and share information in a collaboration and interaction manner. Thus the new form of these technologies changed the teaching and learning paradigm in education. In order to equip students with the necessary skills and knowledge and improve students' engagement, collaborative technologies can be integrated into the learning environment. Therefore, contributions of integrating collaborative technology in engineering education are explained in the following paragraphs.

Students can learn effectively when they actively participate in the learning process. According to Jonassen (1999), knowledge is socially constructed based on a student's interpretation of the world. Therefore, to involve students in the learning process, collaborative technologies are a good option. Collaborative technologies make students more social and engaging (Livingstone, 2015). On the other hand, collaborative technologies provide a learning environment that is socially connected, encouraging collaborative work and promoting the creation of knowledge (Andersen, 2007). In this regard, collaboration can be regarded as an important principle taken into consideration when designing the learning environments of engineering education. Constructivist teaching methods such as problem based learning, cooperative learning, project based learning require collaborative activities. During the collaborative activities, students can both learn the course content effectively and gain the 21st skills that are needed in the business sector after graduation. It can be said that students work in groups and are responsible for their own learning and group members' learning. In addition, students in each group have to do activities such as problem defining, searching for information, proposing solutions, interacting with the necessary resources, communicating with students and instructors for learning. At this point, collaborative technologies can be used and integrated into learning environments. Therefore, a student-centered learning environment can be designed for the active participation of students.

An engineer should have both fundamental skills for the professional engineering life and cognitive abilities that are needed in the changing world. The fundamental skills that an engineer has to gain are mathematics, science, and engineering knowledge. When instructors give lectures to the students in a traditional way, the knowledge passes from instructor to them. This way of teaching is known as the teacher-centered method. Thus, when students graduate from the engineering program, they will have some problems to apply theoretical knowledge to practical life. In order to enhance the engagement of students to the learning process and practice opportunities of theoretical knowledge, hands-on activities with the help of collaborative technologies can be integrated into the learning and teaching process. First of all, hands-on activities are useful in applying theoretical knowledge to practice. These activities lead students to gain both theoretical and practical knowledge by engaging in the learning process. Second, as collaborative technologies allow students to collaborate, to get involved in creating content (Grosseck, 2009), to promote interaction and to share information easily (Franklin & van Harmelen, 2007), they lead to more active learning environments (Richardson, 2009). In other words, while most of the web applications in the web 1.0 era are presentation-oriented, web 2.0 technologies give importance to social connectivity (Ajjan & Hartshorne, 2008). Therefore, students do not only access the course content through web applications but also attend to the collective knowledge construction through social interactions (Maloney, 2007). It can be seen that the positive contribution of hands-on activities for learning new knowledge permanently is being consistent with the features of collaborative technologies. As a result, hands-on activities can be useful to instill engineering students in both theoretical and practical knowledge when the learning environments are supported by collaborative technologies.

An engineer has to be equipped with skills such as problem solving, decision making, critical thinking, communicating, collaborating, team working, self-directed learning (Zhuang & Xu, 2018). Furthermore, according to the Engineer of 2020 Report "employers are looking for engineers with creativity, leadership, entrepreneurial skills, lifelong learning skills, and the ability to work in interdisciplinary teams and to incorporate interdisciplinary knowledge in their work." (p. 75). While there is a demand from the business world, Felder (2012) stated that a graduate student from an engineering program had a low level of critical and analytical thinking, communication and collaboration skills. Therefore active learning techniques can be used in the learning environments. Students can learn by doing such as involving in-class activities, reflecting on their learning, solving a real-world problem, collaborating with other students. As collaborative technologies provide several advantages such as reduction of costs, ease of access to the information (anywhere and anytime), access from several ICT (computer, laptop, tablet or smartphone) with an internet connection, and ease of use (Grosseck, 2009), they can be used to support these techniques.

Integrating Collaborative Technologies into Engineering Education

There are several teaching methods that incorporate collaborative activities. For example, collaborative problem solving, problem based learning, case based learning, cooperative learning, project based learning require collaboration among students and students' involvement in the learning process. In other words, these methods support the notion of social interaction during learning. As characteristics of collaborative technologies are well-suited with the theoretical assumptions about learning of these methods, collaborative technologies can be integrated into the learning environment where one of these methods is being used. However, there are different steps when applying these methods in the courses. Therefore collaborative problem solving method framework (Nelson, 1999) and how collaborative technologies can be used with this method in the learning environment of engineering education are explained in this section.

1. **Build Readiness:** Instructors should start with explaining students about what they will be engaging in and why. Then unstructured problems or real-life problems should be developed for students to solve. When preparing the problems, instructors should be considered the course content, instructional goals, time and characteristics of the students. On the other hand, students'

group process skills should be suitable for collaborative works. In this step, collaborative technologies that are most appropriate can be selected. For example, social network site or blog can be used for the course sharing (Facebook, Edmodo or Blog: Group and personal postings, sharing of collaborative group works, sharing resources), file sharing and storing tool can be selected for the project folder (Google Drive or OneDrive: Collaborative group folder sharing and storing), conferencing tool can be selected for online communication (Google Hangout: Online meetings for group work), document creation tool for preparing documents (Google Documents or Word Online: reporting of collaborative work), mindmapping tools can be preferred for discussing ideas about the problem (Bubbl.us or Mind42: proposing solutions for the unstructured problems) and presentation tools, animation tools, infographic tools, website creation tools, and wiki can be used for creating and presenting the solution. In order to carry out this step successfully, a pilot study can be organized to make students ready for this process. During this study, collaborative technologies should be explained in terms of how they can be used for sharing, collaborating and communicating, how they can be used in which collaborative problem solving step. As a result, the instructor and students build their readiness for collaborative group work.

2. **Form and Norm Groups:** Creating a heterogeneous group is vital. Therefore groups can be formed with 3-6 students according to the criteria such as gender, pre-knowledge and group process skills. Then each collaborative group should prepare their group agreements including group identity, division of labor and how the interaction will be conducted. In this regard, each group can organize an online meeting with conferencing tools and prepare a report about agreements with document creation tools. Then each group create a project folder and share it with group members and instructors via file sharing and storing tools. In addition, virtual collaborative groups can be created on social networking tools.
3. **Determine a Preliminary Problem Definition:** The instructor gives the unstructured problems to the collaborative groups on social networking tools. Then each group should identify the problem. To this end, each group can conduct online meetings and start brainstorming for solutions, the design plan, defining ideas and knowledge, necessary resources. To do this, conferencing tools can be used. On the other hand, document creation tools can be used for preparing the report of the meeting and mindmapping tools can be preferred for the design plan.
4. **Define and Assign Roles:** According to the results of determining a preliminary problem definition, each student should have a role in the collaborative group work. Various roles such as designer, project manager, developer, programmer

or recorder can be selected by group members. In order to do this, conferencing tools and document creation tools can be used. On the other hand, all activities from the starting point to this point can be reported and stored in file sharing and storing tools. Then each group shares their agreements, roles and problem definition and design plan on social networking tools.

5. **Engage in an Iterative Collaborative Problem Solving Process:** Collaborative groups are ready for problem solving process. This step is iterative because groups try to find a solution to the unstructured problem. According to the design plan, groups can engage in activities on a weekly basis. For example, groups can conduct online meetings with group members and instructor with conferencing tool, create report on document creation tools, share what is done for the project and give feedback to the other groups' postings related to project on social networking tools and create a project, solution, product or presentation with other collaborative tools (presentation tools, animation tools, infographic tools, website creation tools, wiki).
6. **Finalize the Solution or Project:** Collaborative groups begin to prepare the draft version of their solution. Therefore each group can share their drafts on social networking tools. Then the instructor and other groups give feedback to the drafts. Finally, each group can revise the draft and prepare the final version of the solution.
7. **Synthesize and Reflect:** After completing the collaborative work, each student can write a reflection about gaining in course content, group process skills, learning strategies, and metacognitive skills on document creation tools and store it on file sharing and storing tools.
8. **Assess Products and Processes:** Each solution, product or project is assessed by the instructor. The instructors should use alternative assessment methods and tools such as rubrics, individual and group reflection reports, portfolios, open-ended exams, and usability tests. At the end of the process, both products and the learning process should be evaluated. In this regard, the quality of the product or solution can be assessed by a rubric and usability test. On the other hand, what the students gain in content knowledge, group process skills and higher order thinking skills can be evaluated with reflection reports, portfolios, and open-ended exams.
9. **Provide Closure**: A closure activity such as celebrating each student can be conducted at the end of the process. This closure activity is necessary for group works. During this activity, efforts for the group work, strong characteristics for group process and experiences gained from each student can be celebrated.

A summary of these steps is presented in Table 1.

Table 1. Collaborative problem solving framework (Nelson, 1999)

• Build Readiness	- Preparing the problems - Selecting the appropriate collaborative technologies o Social network site or blog for course sharing o File sharing and storing tool for a collaborative project o Conferencing tool for online communication o Document creation tool for preparing documents o Mindmapping tools for discussing ideas about the problem o Other tools for creating and presenting the solution - Introducing the process and technologies
• Form and Norm Groups	- Creating collaborative groups - Making online meetings for preparing group agreements (online communication tools) - Creating project folders (file sharing and storing tools) - Assigning groups on social networking tools.
• Determine a Preliminary Problem Definition	- Presenting the problems to the groups (social networking tools) - Conducting online meetings for solution (conferencing tools) - Preparing the meeting report (document creation tools)
• Define and Assign Roles	- Determining each group members' role for group work - Discussing roles (conferencing tools) - Reporting the online meeting (document creating tools) - Storing report (file sharing and storing tools)
• Engage in an Iterative Collaborative Problem Solving Process	- Engaging in activities on a weekly basis according to the design plan.
• Finalize the Solution or Project	- Preparing the draft version of the solution sharing it (social networking tools)
• Synthesize and Reflect	- Writing a reflection report about the process (document creation tools)
• Assess Products and Processes	- Using different tools to assess products and students' gaining.
• Provide Closure	- Conducting a closure activity.

(Source: Nelson, L. M. (1999). Collaborative problem solving. In Reigeluth, C. M. (Ed.), Instructional-Design Theories and Models, Vol. II (pp. 241-269). New Jersey: Lawrence Erlbaum Associates, p. 258)

CONCLUSION AND RECOMMENDATIONS FOR FUTURE WORK

Nature of engineering education and expectations from learners direct instructors and instructional designers to implement hands-on activities, project based learning, and problem based learning strategies in any type of engineering curriculum. Collaborative problem solving method as a general guiding strategy can be used to address the goals of the engineering curriculum and help students to improve their personal development and 21st century learner skills. However, future work is needed to improve the instructional implementations, tools, and evidence of effect in engineering education.

The first area of research is to improve the instructional design models utilized to guide the development of learning environments. Instructional design models guide planning, development, implementation, and evaluation of instruction in learning environments. Currently, the collaborative problem solving model is used in the development of learning environments. However, as in any model of instruction, it lacks providing complete guidance on how to implement and evaluate collaborative problem solving activities. By using formative research methodology, improvements for the collaborative problem solving process can be considered.

The second area of research is related to how to design collaborative learning environments for different areas of engineering education and what type of collaboration tools are needed. Designing instruction is heavily dependent on the context in which the instruction would be implemented. Although there are similarities in curriculum needs and implementation methods, different fields of engineering present different curriculum needs. Therefore, it is important to understand the needs of a particular type of engineering education and develop an effective learning environment that meets these needs. Additionally, different types of engineering education may require a variety of collaboration tools. For example collaboration tools that are currently used in computer engineering education may not meet the needs of electrical or mechanical engineering education. There is a need to develop collaboration tools that serves better for different fields of engineering education. Formative research methodology can be used to improve existing learning environments and collaboration tools or develop new learning environments and collaboration tools to meet the curriculum needs of engineering education.

Finally, more research studies are needed to provide evidence for the effects of collaborative learning activities on students' educational outcomes in engineering education. Implementing collaborative learning activities requires a lot of time and effort on both the instructor and student side. The value of instructional strategies evaluated against the criteria of effectiveness, efficiency, and appealing to student educational outcomes mostly defined as satisfaction, engagement, and achievement. It is important to evaluate the value of collaborative learning activities for student educational outcomes using these criteria over other methods of instruction in engineering education. Experimental or quasi-experimental studies can be used to compare collaborative learning activities and lecturing to understand the effectiveness and efficiency of these methods in terms of student educational outcomes.

REFERENCES

Ajjan, H., & Hartshorne, R. (2008). Investigating faculty decisions to adopt Web 2.0 technologies: Theory and empirical tests. *The Internet and Higher Education*, *11*(2), 71–80. doi:10.1016/j.iheduc.2008.05.002

Albanese, M. A., & Mitchell, S. (1993). Problem-based learning: A review of literature on its outcomes and implementation issues. *Academic Medicine*, *68*(1), 52–81. doi:10.1097/00001888-199301000-00012 PMID:8447896

Albion, P. R. (2008). Web 2.0 in teacher education: Two imperatives for action. *Computers in Schools*, *25*(3-4), 181–198. doi:10.1080/07380560802368173

Andersen, P. (2007). *What is Web 2.0?: ideas, technologies and implications for education*. Bristol: JISC.

Astin, A. (1993). *What Matters in College: Four Critical Years Revisited.* San Francisco: Jossey-Bass.

Bower, M. (2016). Deriving a typology of Web 2.0 learning technologies. *British Journal of Educational Technology*, *47*(4), 763–777. doi:10.1111/bjet.12344

Bower, M., Hedberg, J. G., & Kuswara, A. (2010). A framework for Web 2.0 learning design. *Educational Media International*, *47*(3), 177–198. doi:10.1080/09523987.2010.518811

Butler, J. W. (2012). Grappling with change: Web 2.0 and teacher educators. In D. Polly, C. Mims, & K. Persichitte (Eds.), *Developing Technology-Rich Teacher Education Programs: Key Issues* (pp. 135–150). Hershey, PA: IGI Global. doi:10.4018/978-1-4666-0014-0.ch010

Carini, R. M., Kuh, G. D., & Klein, S. P. (2006). Student engagement and student learning: Testing the linkages. *Research in Higher Education*, *47*(1), 1–32. doi:10.100711162-005-8150-9

Chen, H. L., Lattuca, L. R., & Hamilton, E. R. (2008). Conceptualizing engagement: Contributions of faculty to student engagement in engineering. *Journal of Engineering Education*, *97*(3), 339–353. doi:10.1002/j.2168-9830.2008.tb00983.x

Chen, J., Wang, M., Kirschner, P. A., & Tsai, C. C. (2018). The role of collaboration, computer use, learning environments, and supporting strategies in CSCL: A meta-analysis. *Review of Educational Research*, *88*(6), 799–843. doi:10.3102/0034654318791584

Chickering, A. W., & Gamson, Z. F. (1987). Seven principles for good practice in undergraduate education. *AAHE Bulletin, 3*, 7.

Chis, A. E., Moldovan, A. N., Murphy, L., Pathak, P., & Muntean, C. H. (2018). Investigating flipped classroom and problem-based learning in a Programming Module for Computing Conversion course. *Journal of Educational Technology & Society*, *21*(4), 232–247.

Dolmans, D. H., Loyens, S. M., Marcq, H., & Gijbels, D. (2016). Deep and surface learning in problem-based learning: A review of the literature. *Advances in Health Sciences Education: Theory and Practice*, *21*(5), 1087–1112. doi:10.100710459-015-9645-6 PMID:26563722

Duffy, T. C., & Cunningham, D. D. (1996). *Constructivism: Implications for the design and delivery of instruction. In Handbook of Research for Educational Communications and Technology*. New York: MacMillan Library.

Felder, R. M. (2012). Engineering education: A tale of two paradigms. In B. McCabe, M. Pantazidou, & D. Phillips (Eds.), *Shaking the Foundations of Geo-Engineering Education* (pp. 9–14). Boca Raton, FL: CRC Press. doi:10.1201/b15096-4

Fiore, S. M., Graesser, A., Greiff, S., Griffin, P., Gong, B., Kyllonen, P., ... von Davier, A. (2017). *Collaborative problem solving: Considerations for the national assessment of educational progress*. Retrieved from https://nces.ed.gov/nationsreportcard/pdf/researchcenter/collaborative_problem_solving.pdf

Franklin, T., & Van Harmelen, M. (2007). *Web 2.0 for content for learning and teaching in higher education*. Retrieved from https://staff.blog.ui.ac.id/harrybs/files/2008/10/web-2-for-content-for-learning-and-teaching-in-higher-education.pdf

Greenhow, C., Robelia, B., & Hughes, J. E. (2009). Learning, teaching, and scholarship in a digital age: Web 2.0 and classroom research: What path should we take now? *Educational Researcher*, *38*(4), 246–259. doi:10.3102/0013189X09336671

Grosseck, G. (2009). To use or not to use web 2.0 in higher education? *Procedia: Social and Behavioral Sciences*, *1*(1), 478–482. doi:10.1016/j.sbspro.2009.01.087

Jonassen, D. H. (1999). Designing constructivist learning environments. In C. M. Reigeluth (Ed.), *Instructional-Design Theories and Models* (Vol. 2, pp. 215–239). Lawrence Erlbaum Associates.

Korucu, A. T., & Cakir, H. (2018). The effect of dynamic web technologies on student academic achievement in problem-based collaborative learning environment. *Malaysian Online Journal of Educational Technology*, *6*(1), 92–108.

Kuh, G. D., Cruce, T. M., Shoup, R., Kinzie, J., & Gonyea, R. M. (2008). Unmasking the effects of student engagement on first-year college grades and persistence. *The Journal of Higher Education, 79*(5), 540–563. doi:10.1080/00221546.2008.11772116

Laird, T. F. N., & Kuh, G. D. (2005). Student experiences with information technology and their relationship to other aspects of student engagement. *Research in Higher Education, 46*(2), 211–233. doi:10.100711162-004-1600-y

Livingstone, K. A. (2015). *The impact of Web 2.0 in Education and its potential for language learning and teaching*. Retrieved from https://www.researchgate.net/publication/272487634_The_impact_of_Web_20_in_Education_and_its_potential_for_language_learning_and_teaching

Maloney, E. (2007). What Web 2.0 can teach us about learning. *The Chronicle of Higher Education, 25*(18), B26.

National Academy of Engineering. (2004). *The engineer of 2020*. Washington, DC: National Academies Press.

Nelson, L. M. (1998). *Collaborative problem solving: An instructional theory for learning through small group interaction* (Unpublished doctoral dissertation). Indiana University.

Nelson, L. M. (1999). Collaborative problem solving. In C. M. Reigeluth (Ed.), *Instructional-Design Theories and Models* (Vol. 2, pp. 241–269). Lawrence Erlbaum Associates.

Norman, G. R., & Schmidt, H. G. (1992). The psychological basis of problem-based learning: A review of the evidence. *Academic Medicine, 67*(9), 557–565. doi:10.1097/00001888-199209000-00002 PMID:1520409

O'Reilly, T. (2005). *What is web 2.0: Design patterns and business models for the next generation of software.* Retrieved 9 from http://oreilly.com/web2/archive/what-is-web-20.html

Ohland, M. W., Sheppard, S. D., Lichtenstein, G., Eris, O., Chachra, D., & Layton, R. A. (2008). Persistence, engagement, and migration in engineering programs. *Journal of Engineering Education, 97*(3), 259–278. doi:10.1002/j.2168-9830.2008.tb00978.x

Ravitz, J. (2009). Introduction: Summarizing Findings and Looking Ahead to a New Generation of PBL Research. *Interdisciplinary Journal of Problem-Based Learning, 3*(1).

Richardson, W. (2009). *Blogs, wikis, podcasts, and other powerful web tools for classrooms* (2nd ed.). Thousand Oaks, CA: Corwin Press.

Rotgans, J. I., & Schmidt, H. G. (2011). Cognitive engagement in the problem-based learning classroom. *Advances in Health Sciences Education: Theory and Practice, 16*(4), 465–479. doi:10.100710459-011-9272-9 PMID:21243425

Şendağ, S., & Odabaşı, H. F. (2009). Effects of an online problem based learning course on content knowledge acquisition and critical thinking skills. *Computers & Education, 53*(1), 132–141. doi:10.1016/j.compedu.2009.01.008

Tinto, V. (1993). *Leaving college: Rethinking the causes and cures of student attrition*. Chicago, IL: The University of Chicago Press.

Topalli, D., & Cagiltay, N. E. (2018). Improving programming skills in engineering education through problem-based game projects with Scratch. *Computers & Education, 120*, 64–74. doi:10.1016/j.compedu.2018.01.011

Vygotsky, L. S. (1980). *Mind in society: The development of higher psychological processes*. Harvard University Press. doi:10.2307/j.ctvjf9vz4

Yew, E. H., & Goh, K. (2016). Problem-based learning: An overview of its process and impact on learning. *Health Profession Education, 2*(2), 75–79. doi:10.1016/j.hpe.2016.01.004

Zhuang, T., & Xu, X. (2018). New engineering education in Chinese higher education: Prospects and challenges. *Tuning Journal for Higher Education, 6*(1), 69–109. doi:10.18543/tjhe-6(1)-2018pp69-109

ADDITIONAL READINGS

Astin, A. W. (1997). *What matters in college?* JB.

Bower, M., Hedberg, J. G., & Kuswara, A. (2010). A framework for Web 2.0 learning design. *Educational Media International, 47*(3), 177–198. doi:10.1080/09523987.2010.518811

Chickering, A. W., & Gamson, Z. F. (1987). Seven principles for good practice in undergraduate education. *AAHE Bulletin, 3*, 7.

Felder, R. M. (2012). Engineering education: A tale of two paradigms. In B. McCabe, M. Pantazidou, & D. Phillips (Eds.), *Shaking the foundations of Geo-Engineering education* (pp. 9–14). Boca Raton, FL: CRC Press. doi:10.1201/b15096-4

Ohland, M. W., Sheppard, S. D., Lichtenstein, G., Eris, O., Chachra, D., & Layton, R. A. (2008). Persistence, engagement, and migration in engineering programs. *Journal of Engineering Education*, *97*(3), 259–278. doi:10.1002/j.2168-9830.2008.tb00978.x

KEY TERMS AND DEFINITIONS

21st Century Skills: A general term that refers to competencies such as problem solving, critical thinking, collaboration.

Collaboration: Making decisions and taking actions based on a group of people's common understanding of any problem at hand with shared goals and values.

Collaborative Problem Solving: An instructional approach that students work in a group, engage in problem solving activities, and develop 21st skills during the process.

Collaborative Technologies: Collaborative technologies are web applications that allow users to create, analyze, exchange, and share information in a collaboration and interaction manner.

Engagement: Level of students' participation in the educational activities in and outside the classroom.

Learning (Social Constructivist): Individuals create meaning of reality by communicating and interacting with each other meaning. The truth generated by reaching consensus is relative to the society in which it is created.

Student-Centered Learning Environment: A learning environment where students can participate in the learning process actively.

Chapter 7
Challenge-Based Learning:
A Multidisciplinary Teaching and Learning Approach in the Digital Era – UoL4.0 Challenge: A CBL Implementation

Eliseo L. Vilalta-perdomo
University of Lincoln, UK

Rosario Michel-Villarreal
University of Lincoln, UK

Geeta Lakshmi
https://orcid.org/0000-0002-8351-719X
University of Lincoln, UK

Chang Ge
University of Lincoln, UK

ABSTRACT

This chapter illustrates a research focused on how to effectively implement the challenge-based learning (CBL) approach in a higher education institution (HEI) in the UK. The challenge was linked to contemporary research conducted by a group of academics, which concerned how digital technologies can positively impact the local economy. The project was named 'UoL4.0 Challenge', and it proved that designing and implementing CBL educational environments can increase students' propensity to work actively and proactively. The exercise also suggested that CBL may support students in the application of their academic skills and digital capabilities to support their communities. This study presents a description of the case and a reflection on lessons learned with an aim to provide guidelines for other educators and policymakers that are interested in implementing I4.0 educational initiatives at local or national levels. It is also suggested that CBL may play a fundamental role in implementing the triple-helix model of innovation.
DOI: 10.4018/978-1-7998-2562-3.ch007

INTRODUCTION

The World Development report (2019) on *The Changing Nature of Work* delineates some of the impacts that new digital technologies will have on working environments. For instance, many low-skill jobs are at risk, as robots become cheaper and more effective in dealing with routine tasks; but new work opportunities also emerge. Digital technologies linked to day-to-day operations provide massive data collections and reliable data analyses, available at the click of a button for anybody with access to a personal computer. Consequently, better-informed decision-making has become a reality for many managers, even in micro-businesses that traditionally suffer from limited resource availability. However, the most outstanding insight that the World Bank report provides is the confirmation of the primacy of human capital; workers can be supported but not substituted, by robots and artificial intelligence. Accordingly, utopian expectations of humans being served by herds of mechanical slaves (robots) seem a far-fetched dream. Following this line of argument, the World Bank claims that one imminent global challenge involves how to prepare future generations for a different labor market. Some argue, as the World Bank in their 2019 report, the need to solve this by focusing on children currently in primary schools. Nevertheless, this may not necessarily the best strategy, as it seems impossible to foresee what our children will confront in the future and thus to teach them precisely those skills. Interestingly enough, what seems to be commonly recognized is that to be effective in a digitalized world, national education systems must provide "a combination of technological know-how, problem-solving, and critical thinking, as well as soft skills such as perseverance, collaboration, and empathy" (World Bank, 2019).

Different countries recognize the impact that digital technologies are having in their economic and social development. Their governments are taking steps towards such technology integration, through comprehensive policies concerning digital technologies. It was the German government who coined 'Industrie 4.0' (I4.0) as the umbrella "to drive digital manufacturing forward by increasing digitalization and the interconnection of products, value chains and business models" (EC, 2017), but others soon followed. For instance, the Italian government has also presented a national plan for I4.0 (MdSE, 2019). The Italian initiative follows three main principles: (a) to operate in a logic of technological neutrality, (b) to intervene with horizontal and not vertical or sectoral actions, and (c) to act on enabling factors. Many of the European efforts in this field are aligned to the EU concept of 'Factory of the Future' (EC, 2018). Outside of Europe, other governmental actors can also be found, such as the Mexican government who has developed a strategy based on four pillars: (a) the development of human capital, (b) innovation, (c) clusters and (d) technology adoption to support I4.0 implementation inside their economy. Accordingly, the Mexican Ministry of Economy has delineated an I4.0 RoadMap

(ProMexico, 2017). Other governmental actors have even developed norms to describe and regulate I4.0 activities. For instance, the Spanish Association for Standards and Certification (Asociación Española de Normalización y Certificación, AENOR) defines the requirements for managing an I4.0 digitization system (UNE, 2018).

In the UK, the Industrial Strategy Green Paper (2017) links I4.0 to the development of their workforce skills. Inside England, at the regional level, Greater Lincolnshire recognizes that more needs to be done to prepare current and future workforce for the challenges and opportunities of digitalization. The skills identified go beyond the technical one, and include softer ones, such as creativity, curiosity, ability to work across disciplinary boundaries, leadership and vision. What seems to have been given less attention is the development and transfer of industrial digitalization skills. Specialized companies are working on effective and efficient last-mile delivery of products, but this is limited to products with prices able to cover logistics costs. Such business models do not consider SMEs, as their earnings are usually too low to absorb technology costs linked to I4.0 initiatives.

In this context, the University of Lincoln (UoL) recognizes the importance of the digital component in a modern economy and have engaged in a series of initiatives concerning digital technologies and their implementation. An example of the kind of initiatives that UoL is implementing to integrate I4.0 principles into their curricula, is the 'UoL4.0 Challenge'. This is an educational initiative where multidisciplinary teams of students work closely with small- and medium-sized Lincolnshire's businesses, to investigate the level of readiness of Lincolnshire businesses for adopting Industry 4.0 principles and practices. Accordingly, the UoL4.0 Challenge expectations are twofold. For students, the project aims at increasing their employability, by means of developing technology-based entrepreneurial and problem-solving skills, to better deal with cyber-physical systems. For businesses, the project aims at connecting their new commercial ideas with their potential markets, through the development and use of digital technologies.

The authors were closely involved in designing and testing this research project in two ways. Firstly, we explored whether Challenge Based Learning (CBL) was a plausible cross-disciplinary teaching strategy in University of Lincoln in 2018. Secondly, we tested CBL as an effective learning experience for inter-disciplinary UoL students, through the execution of the UoL4.0 Challenge 2019. The aim of our project was to broaden, develop and enrich teaching and learning tools at UoL.

In this chapter we describe, reflect and report on the use and implementation of Challenge-Based Learning approach inside UoL, as a way to develop digital literacies in students to enable them to act confidently in the digital age, and to support local businesses in their exploration of new commercial ideas concerning the use of digital technologies. This initiative known as UoL4.0 Challenge concerns how digital technologies can positively impact local economy. Moreover, we discuss future directions and the impact of the Challenge.

The outcome of our initiative showed that the design and implementation of CBL educational environments can increase students' propensity to work actively and proactively, and to apply academic skills and digital capabilities in enriching personal experiences (Johnson et al., 2009). A description of the case and a reflection on lessons learned provide effective guidelines for other educators and policymakers interested in implementing I4.0 educational initiatives at the local or national levels.

In the next section, we present a brief background of where our activity took place and details of the principles of the CBL framework. This is followed by a detailed account of UOL4.0 planning and implementation, the lessons learnt including the impact and finally we conclude.

BACKGROUND

The University of Lincoln (UoL) is a public university located in the United Kingdom. It is one of UK newest universities that began operation in 1996 (formerly known as University of Lincolnshire and Humberside). Currently, UoL has more than 15 thousand students enrolled in 24 Schools within their 4 Colleges, and addresses research challenges mainly in areas such as agri-food technology, creativity, digital arts, digital archivy, personalized health and rural communities. Its impact is worldwide as the global alumni community consists of more than 90 thousand graduates across 135 countries. The UoL teaching and learning capabilities have been recognized with the Gold category in the Teaching Excellence Framework and its position in national and international rankings has improved consistently. UoL core principles are based on: (a) resilience and sustainability, (b) ambition, global recognition and growing their reputation for innovation, (c) creating an inclusive community, (d) enabling technologies, excellent research and teaching spaces, and (e) a new vision for education (UoL, 2016). These principles are highlighted in different UoL educational initiatives; for instance, their commitment to PRME principles, the "Student as Producer" initiative and new forms of educating, such as Challenge-Based Learning. In a nutshell, UoL is committed to contributing to solving global challenges through the development of local solutions. For instance, recently, UoL led a collaborative bid across Greater Lincolnshire to establish the Lincolnshire Institute of Technology, the focus being to develop technology in agri-tech and food manufacturing, energy and engineering, with a unifying digital theme. In order to align their educational aims and efforts to what is required by the community; UoL has also implemented a multi-actor initiative, concerning the integration of industrial digitalization principles into their academic curricula. Businesses and academics have been working in the "co-creation of a suite of modules and curricula to equip our graduates and workplace learners to meet the digitalization challenges that our industrial partners face now and in the future." This project was funded by the former Higher Education Funding Council for England (HEFCE), currently known as the Office for Students (OfS).

UOL4.0 has its roots in the theory of Challenge-Based Learning. CBL is "an engaging multidisciplinary approach to teaching and learning that encourages students to leverage the technology they use in their daily lives to solve real-world problems" (Apple, 2010) and is part of a larger collaborative project known as *Apple Classrooms of Tomorrow – Today* ($ACOT^2$). This is an initiative created in 1985 for public schools, universities, research agencies, and Apple Computer, Inc. to collaborate in introducing technology inside classrooms as a tool for learning and a medium for thinking, collaborating, and communicating (Ringstaff & Yocam, 1996). CBL is a teaching model that aims at integrating successful educational practices, such as problem-based learning, project-based learning, and contextual teaching and learning, to capture students' engagement (Johnson et al., 2009).

As a relatively new approach, Challenge-Based Learning (CBL) has recently attracted attention from educators as an approach to address the issues explained previously. The approach advocates the use of technology, teamwork, self-directed learning, peer learning, real-world problem solving and reflective learning (Johnson et al., 2009) to acquire also transversal skills, such as critical thinking, problem solving and collaborative work. Previous CBL experiences show that when students realize that their education is preparing them for that, their commitment and engagement increases drastically (Tecnológico de Monterrey, 2015).

The CBL Framework consists on a series of steps that cascade from an initial Big Idea, into "an essential question, a challenge, guiding questions, activities, resources, determining and articulating a solution, taking action by implementing the solution, reflection, assessment, and publishing" (Apple, 2010). These activities also include sharing findings with the external community, by means of videos and other kinds of publications (Dutra Moresi et al., 2017). According to Apple (2010) the key components in the CBL process are:

- Big Idea is a wide concept that encapsulates an area of interest for students that is also pertinent to their communities. A big idea must be able to trigger students' commitment. Hence, it should be capable of being explored from many different perspectives to capture students' different interests. In this sense, big ideas are not just questions bounded through specific sets of parameters, such as a problem that is expected to be solved. The main source for inspiring big ideas rests in the intersections between students' interests and their community needs.

- Essential question is the articulation of the big idea. It shall incorporate a set of statements in the form of questions that capture different students' interests and community needs. Essential questions provide contextual information and inform students what is important to know about the big idea.
- The challenge translates each essential question into a specific task for students. The aim of the challenge is to provide students with insights about which is the kind of answer or solution that might be turned into a sensible action. This challenge needs to be built in such a way that interest and commitment is maintained by students and the organizations involved.
- Guiding questions, usually generated by students, trace a discovery trajectory to indicate what needs to be known and researched to succeed in dealing with the challenge.
- Guiding activities are developed and made available to students to support their research voyage. They may follow one of many problem-solving frameworks available. It also provides students with a sense of timing and urgency, to avoid undesirable side-effects such as 'paralysis by analysis'.
- Guiding resources are sets of resources available to students, which support their research effort. These resources may take different formats, such as digital (podcast and websites), face to face (interactions with experts and organizations) or more traditional (academic and professional publications).
- Solutions in different formats might be produced by creative students, so an open-access repository needs to be considered, for solutions to become accessible to a wider audience. This repository shall provide students with space to experiment their creativity in a safe environment. Recommended formats are those that support dissemination of what was developed during the challenge. For instance, podcasts, short videos or posters could be effective ways to disseminate the proposed solutions.
- Assessment concerns a constructive feedback that sits on a set of criteria to define the level success. There are different ways of assessing the final products, but it is important for students and organizations to know *a priori* what the evaluation rubric will be.
- Publishing aims at making accessible to a wider audience of what has been learnt. It is a way of democratizing knowledge by making accessible to as many as possible.

In order to implement the CBL framework, Apple (2010) proposes a process to mirror the 21st century workplace. This process is a starting point and is not meant to be prescriptive.

- Setting up a collaborative environment. The environment where students and organizations will interact is of utmost importance. To replicate a 21st century workplace, the impact of automation needs to be considered, as it is reshaping work and the skills required. According to the World Bank report (2019), advanced cognitive skills, such as critical thinking and problem-solving, and socio-behavioral skills, like creativity and curiosity, are in major demand, whilst job-specific skills are declining. Furthermore, synchronous and asynchronous modes of work need to be replicated, with digital platforms accessible in a 24/7/365 mode. Space for face to face interaction is also important as interaction skills, such as empathy, need also to be developed.
- Introduction. It is important to provide a clear introduction about the big idea that will be explored and the challenges that will be developed. Students need to understand what is expected from them, and organizations involved must be aware of the boundaries of the students' intervention.
- Team formation. There are many discussions on how to form groups in educational environments. Even though it is not important how the groups emerge, it is fundamental to consider the roles and responsibilities, and the abilities that each member may add to the team.
- Assessment. The measure of success needs to be established from the beginning. Students and organizations participating in the challenge must understand the project rubric to recognize the success of the solution proposed.
- Guiding questions. After teams are formed, students begin the research process by identifying what they know and what they need to know. If this becomes difficult for students to start this process, a possibility is the teaching team to support the beginning of this process by providing some initial questions.
- Guiding activities and guiding resources. A series of guiding activities, supported by guiding resources, can be designed to orient the teams in the research process. One approach to provide effective support is to provide students with a problem-solving framework that will help them follow a systematic research process. The aim of this step is that by following the guiding activities and resources, students will be supported in the development of a series of solutions, and to evaluate which of them seems better, in terms of its efficacy, efficiency and effectiveness.
- Prototype / testing. After a set of solutions has been identified by the team, further refinement is achieved by building prototypes and testing their performance, by implementing each solution within small groups or by presenting them to a focus group for deeper discussion.
- Implementation. Once one specific solution has been considered as the most appropriate to solve the challenge, it is the time to implement the solution. This implementation is expected when the students' team group has enough

time to do such implementation and evaluating the outcomes. In some cases, additional time and resources should be available, as putting the plan "in a real-life setting is important" (Apple, 2010).

- Assess. The project rubric is used to identify the quality of the solution provided. This helps students to reflect on how to improve their individual and collective skills in future events.
- Reflection / documentation. CBL is a learning process, where self-reflection plays a fundamental role. Students should document, step by step, what they do individually and collectively and why. Documenting provides evidences for a process self-reflection, where students become aware of what they have learnt about themselves, working with others, the challenge, potential solutions to such challenge and how to solve future challenges.
- Publish. Students are asked to publish their work in different formats and locations. Different formats may be used, so students should be supported in increasing their abilities on communicating their ideas to a variety of audiences. Videos, posters, reports, blogs and other media should be considered for the challenge.
- Ongoing informative assessment. Finally, as any learning activity, feedback is required to achieve closure. Students should be informed and made aware on how they progressed in terms of their individual and collective skills. This can be done by means of formative or summative assessment, or a combination of both.

In summary, CBL learning has been recognized as an educational process that increases, at least, four competences: (a) problem resolution; (b) leadership, entrepreneurship and innovation; (c) ethics, and (d) collaborative work (Ramirez-Mendoza, et al., 2018). According to Johnson et al. (2009), giving the opportunity to participate in a CBL results in students with increased engagement, more time spent, creative application of technology, and increased student satisfaction.

UoI4.0 Challenge – Its Design and Execution

World-class universities are recognized for their intellectual contributions, particularly in knowledge-based economies. The importance for universities to develop knowledge and to become imparters of education in how to produce such knowledge demands, involves positioning Higher Education Institutions (HEIs) as more than training centers, which focus on fulfilling specific demands of specialized labor markets. Research processes must embed university tasks; consequently, how to practice research must form integral teaching practice. One way to fulfil this responsibility of creating knowledge and teaching this as a skill, is allowing students to actively participate in ongoing research, as a means to acquire the ability to conduct their own research processes in their future professional lives. Hence Research-Based Learning is a proven educational technique to effectively incorporate research into undergraduate programs (Morales-Menendez, et al., 2019).

The University of Lincoln (UoL) has taken this approach as fundamental part of their educational ethos, through the 'Student as Producer' initiative. This initiative reconnects core activities of universities, research and teaching, and emphasizes the collaborator role that students may play in the production of knowledge (University of Lincoln, 2013). A further development at UoL, concerning the integration of research skills into different programs, involves Challenge-Based Learning (CBL). In the following section we present a two-year project to incorporate CBL in the educational toolbox at UoL. Within UoL this project is known as the 'UoL4.0 Challenge'.

Describing the Start of UoL4.0 Challenge

In 2017, a group of academics at University of Lincoln (UoL0) indicated a common interest in enhancing the curricula with skills associated to digital industrialization, what is also known as Industry 4.0 (I4.0). Two of them had previous experiences in CBL as a teaching and learning approach, and they proposed other colleagues to participate in a pilot experience. The plan was to break the project into two parts. The first one, focused on designing a CBL intervention that incorporated digital skills, was named 'UoL4.0 Initiative'. The second one was a pilot to execute the plan and test the design principles.

Stage 1. UoL4.0 Initiative – Designing the challenge

The Lincoln International Business School (LIBS), part of the University of Lincoln (UoL), has a small seed program to fund innovative teaching and learning initiatives. The academic group decided to benefit from this funding to support a CBL design process. The trigger for taking advantage from this fund was exploring the impact of digital technologies in Lincolnshire economy. Accordingly, a proposal to develop a Challenge-Based Learning approach for connecting business ideas to the market was prepared. The aim of such proposal was to put Industry 4.0 opportunities within the reach of the UoL community, mainly students and associated micro- and small entrepreneurs.

The goals for the UoL4.0 Initiative were to: (a) increase the awareness of I4.0 inside the UoL community (internal and external); (b) build capacity for business/ practical implementations of I4.0 within the UoL community (internal and external); (c) support the design and implementation of I4.0, particularly at the micro-level, and (d) reflect on implementations conducted and disseminate lessons learnt. The expected outcomes were to: (a) trigger the interest of the UoL community on technology-based entrepreneurial activities; (b) develop an internationally recognized differentiator, centered on an incipient technology and business opportunity; (c) act as a strong regional actor on developing and supporting innovative technology-based micro-businesses, and (d) strengthen the exit toolbox for UoL graduates interested in the technology-based economy.

In the stage 1 of this project, six UoL academics were invited to participate, from the Business School (4) and the School of Computing (2) along with three international academics from Mexico, Peru and Spain. Some of these participants were more involved than others.

Stage 1 of the project focused on increasing the awareness of I4.0 inside the UoL community. The main idea was to design a 'Challenge-Based Learning' (CBL) initiative that supports the development of highly engaged, employable and creative-thinking graduates who contribute to the development of society and the economy. The UoL4.0 Initiative targeted students interested in developing their technology-based entrepreneurial skills, particularly cyber-physical systems.

The preliminary plan of activities to be conducted within the UoL4.0 Initiative are presented in table 1.

Table 1. UoL4.0 initiative. activities according to the kind of action

Expertise building	Expertise dissemination	Organizational activities
1. Identify UoL partners involved/ interested in I4.0 projects. E.g. Siemens. 2. Invite speakers with knowledge on I4.0. E.g. CILT, Siemens, other universities, government officials, Erasmus+. 3. Associate I4.0 initiative to particular modules. E.g. start-ups, logistics and operations management, commercial and operational management, or entrepreneurship. 4. Design a competition for UoL students on I4.0 initiatives. To link this competition with external agents, such as Siemens or the Lincoln Business Improvement Group. Even a sandpit exercise was initially considered	1. Design and maintain a repository (e.g. Wordpress©). 2. Feed the repository with relevant didactic material on I4.0 and micro-businesses. 3. Build a distribution list.	1. Approach potential companies to participate/sponsor the challenge. 2. Identify current level of awareness about I4.0 among our students. 3. Develop a 'big idea' and 'essential question' that could be used for a CBL concerning I4.0. 4. Design the CBL exercise involving I4.0. 5. Prepare a series of guiding questions, activities and resources, and an assessment system for the CBL.

In practice, difficulties arose affecting the plan, so adaptations were made. For instance, the initial idea was to conduct a survey to identify the current I4.0 awareness of students. However, internal conversations within the academic team developed an alternative that would provide more information to the academic design team. Students from different Colleges (Arts, LIBS and Science) were invited to participate actively, rather than passively through the survey, in the conceptual development of UoL4.0 Design, by means of a focus group.

The focus group was run with twelve students, two of these from one BSc program and the rest from six different MSc programs. The problem analyzed during this focus group was to reflect if CBL could be used to explore how to connect products and services with the market, through digital technologies and business solutions. The specific Big Idea proposed to the students was 'how digital technologies may impact positively Lincolnshire's economy?' The activities during the focus group involved: (a) making a short presentation by academics to explain the concept of Industry 4.0; (b) making another short presentation on the understanding of Challenge-Based Learning; (c) initiating a group activity to move from the Big Idea to the Essential Question, and (d) running a final group activity where the principles behind the organisation of the event (date, duration, support materials, etc.) were discussed and proposals from students were received.

In terms of outcomes, the students' reception to the idea was very positive, and their interest in participating in a future event was explicit and promising. Students were able to recognize different proposals, concerning areas of interest, such as agriculture, automotive industry, government (NHS + Police), and tourist industry. This indicated academics could be confident for a quick and successful transfer of the concept of I4.0 to students with no previous knowledge. The group of students also proposed an interesting future research challenge: how does a company evaluate its current digital maturity?

The initial structure proposed by the academic team was to run one 4-day event, from Monday to Thursday. However, students indicated a preference to incorporate a break in the middle, rather than having all the days together. This preference informed the following year's UoL4.0 Challenge in 2019, where the event ran from Thursday to Tuesday, with a weekend in the middle. Finally, students asked for additional support; in particular, experts' talks, materials and space where to work together. These too were embedded in the design of the UoL4.0 Challenge 2019.

The CBL concept, too, was positively received by local businesses. The initial idea was to identify one organization acting as participant/sponsor, as this simplifies the design of the challenge and the control and monitoring of the event. However, after discussing this possibility with different organizations, the positive response allowed us to widen the activities to four companies: an agri-food company, a theatre, a marketing and publisher organization, and an office within the University of Lincoln. Additionally, other explorations with three potential sponsors took place; as a result, the Chartered Institute of Logistics and Transport (CILT) supported the award ceremony at the end of the event.

Academics from different departments at LIBS and other two schools (Computing and Design) were included during the stage; some of them continued to participate actively in this initiative. The project was assessed by an external academic specialist in the field, from the Spanish *Universidad International de la Rioja* (UNIR). The assessment provided a typology of different ways that a UoL4.0 CBL may be designed and conducted. The selected version for a future edition of the UoL4.0 Challenge considers "making a comparison of tools available in the market to solve a problem that requires Digital Transformation" (Rainer-Granados, 2018).

Concerning dissemination of the outcomes of this first stage of the UoL4.0 initiative, the UoL4.0 Design, plans included developing papers for two different conferences but no funds were available from the research budget, so dissemination had to adjust; so, the academic team made good use of the opportunities available. Externally, the dissemination delivered was: (a) a presentation for members of the Royal Logistics Corp at the EY HQ in London; (b) a presentation to members of the East Lindsay businesses group at Mortons Media Group, and (c) participation with a stand at the #GoDigital18 conference. Internally, there was a presentation at the Lincoln International Business School (LIBS) Teaching and Learning Conference 2018. This presentation was a summary of the UoL4.0 initiative. As a way to explore different formats for presenting that students might follow in a future event, this presentation followed the challenging PechaKucha format. This presentation had a warming reception from LIBS colleagues.

In summary, this first stage of the UoL4.0 Initiative, indicated the need to develop a CBL event that would consider: (a) students teamwork, (b) working around the clock, (c) one week, (d) in a real solution, (e) for a real technology-based problem, and (f) as the vehicle for an enriching T&L experience.

Stage 2. UoL4.0 Initiative – Executing the Challenge

As indicated above, the UoL4.0 Challenge initiative is based on the 'Challenge-Based Learning' educational strategy to support students and Lincolnshire businesses and organizations in the discovery of opportunities concerning the digitalization of industry and services (e.g. Big Data/Analytics, Blockchain, the Cloud, Internet of Things and wearable sensors). UoL4.0 Challenge innovation lies in a collective action that supports aims, preferences and expectations from different organizational actors interested in exploring their current and future business models validity. Accordingly, the UoL4.0 Challenge 2019 Initiative goals were threefold and aimed at:

- Students, who will increase their employability levels by means of developing technology-based entrepreneurial and problem-solving skills, to better deal with challenges posed by cyber-physical systems.
- Businesses, who support students' teams, will explore how to connect their new commercial ideas with their potential markets, by means of developing and using digital technologies and applications.
- Academics, who support students' teams, will be able to develop and test Challenge-based learning principles and examples that may be disseminated to other I4.0 communities around the world.

As previously indicated, the UoL4.0 Challenge project was of a research nature, and its goals were twofold. First, to investigate the use of Challenge-Based Learning (CBL) as an innovative and effective cross-disciplinary teaching strategy at the University of Lincoln. Second, to explore the impact of the learning experience derived from the UoL4.0 Challenge 2019 initiative in different UoL students. In summary, this project aimed at enriching UoL Teaching and Learning Toolbox, with an innovative approach such as CBL. Accordingly, the research tasks for the UoL4.0 Challenge 2019 were: (a) to conduct a challenge-Based learning intervention at the University of Lincoln, and (b) to assess the impact of challenge-based learning on students' learning experiences through thematic analysis of reflective accounts and survey. The research questions guiding the investigation were as follows: (a) What are the first-hand lessons collected from the use of Challenge-Based Learning (CBL) as an innovative and effective cross-disciplinary teaching strategy at the University of Lincoln?, (b) What is the impact of a challenge-based learning intervention on students' learning experiences?

UoL4.0 Challenge 2019 took place during six days in February 2019. The first day brought together students, companies and academics to have initial discussions on the challenges. The following days, students worked in teams with representatives of the companies to develop solutions to the challenges. Experts delivered several workshops, covering key themes regarding the development of original ideas and effective engagement with businesses in the digital industrial revolution. An award ceremony provided students the opportunity to communicate their ideas in a business environment.

In the UoL4.0 Challenge 2019 edition, organizations were selected from diverse industry sectors and were of different sizes, with the aim of illustrating how principles of Industry 4.0 may be implemented in any kind of industry and in any size of company. These organizations were:

- An agri-food business. The challenge concerned the need to characterize the dry matter of all the potatoes in a batch, which is impossible to do by analyzing samples. This limitation affects the quality of finished products, as the amount of dry matter may differ. In order to increase uniformity in the quality of finished products, there is the need to measure the dry matter of every single potato in the processing line. With this in mind, a series of questions were provided to the team of students: Which digital technologies are available to analyze the amount of dry matter of all the potatoes in a batch in the processing line (i.e. potatoes grading)? Which data can be collected through such digital technology? What knowledge may be extracted from such data?

- A creative industry organization. This Theatre is a Grade II listed heritage building. Activities relating to the Theatre are concentrated in the later hours of the day. Accordingly, the challenge focused on how to increase the amount of productive activities within the Theatre out of the later hours of the days. The main question presented to students concerned the use of digital technologies to help increase the utilization rate of the building.
- A marketing company. This organization holds a valuable reservoir of information concerning potential customers. However, there is no workforce allocated to analyze such database. The challenge focused on how to transform the available data into accessible information that would be used to increase sales. The challenge lies in the limited amount of human resources, so the data analysis should be conducted without too much human participation. The question was related to the use of digital technologies to provide effective and efficient customers' data analysis.

A team of students was assigned to each organization. During six days of contact with their organization, students worked in developing exploratory ideas around the use of digital technologies to provide workable solutions for each challenge. These solutions involved in the final propositions were: (a) for the agri-food business, the use of Near-Infrared spectroscopy (NIR) to analyze the amount of dry matter; (b) for the Theatre, the use of Projection Mapping for any surface inside the Theatre to become a canvas for any means of animations, video and imagery, and (c) for the marketing business, the use of Hypertext Preprocessor (PHP) to analyze data on what customers are buying, but not with the company; this to identify which are other products bought by customers, to reflect on why these are bought somewhere else, and to strategize on how these products could be also sold by the company.

Lessons Learned

Traditional project-oriented learning approaches concern designing problems to be solved under specific topic constraints. Conversely, Challenge-Based Learning

(CBL) is a framework where students, teachers and community members address local and global Challenges while acquiring knowledge through collaboration. By means of CBL, students and teachers can positively impact their communities and demonstrate that learning can be deep, engaging, meaningful and purposeful. UoL4.0 Challenge 2019 was a CBL focused on investigating how digital technologies may support Lincolnshire's economic development. It, thus, served as a bridge between the academic/expert skillset and the possible problems faced in the business world, exposing all parties to be creative in their thinking. In relation to the schedule for the project, table 2 shows the level of flexibility and adjustment that was required.

Table 2. Major tasks and milestones delivered in this project

Original Tasks and Milestones	Original Timeline	Variations / additional deliverables
Meeting with all participants businesses	Friday, 16th of November 2018	On time
Approved challenges' briefings	Friday, 7th of December 2018	On time
Webpage (OneNote) ready for event	Friday, 14th of December 2018	End of January 2019
Event final preparation and organisation	Thursday, 31st of January 2019	On time, apart from finding spaces for group work, that never worked well.
Materials acquisition	Thursday, 31st of January 2019	On time
Students' initial engagement	Friday, 15th of February 2019	On time
Event execution and data collection	Thursday 21st to Tuesday 26th of February 2019	On time
Awards ceremony	Wednesday, 27th of February 2019	Delayed to 4th of March due to conflicting agendas
Analysis of data collection from survey	Friday, 29th of March 2019	Delayed to 12th of April. Some students didn't answer the survey
Analysis of data collection from thematic analysis	Friday, 26th of April 2019	On time
Internal reflection meetings on the analyses results	Four weekly meetings during May 2019	On time. Only one debriefing meeting was required.
Dissemination	End of June	Different dissemination events attended. Additional dissemination work in course.

In this context, the most outstanding effect of UoL4.0 Challenge 2019 was that the main beneficiaries were those who participated directly in it, students and organizations. Specific technological answers to the organizational challenges presented were provided by teams of students working together and offered the following immediate benefits:

1. For students:
    a. to develop and demonstrate digital capabilities and transferable skills to manage real Industry 4.0 business projects;
    b. to learn how to pitch ideas in a business environment, and
    c. to improve their potential employability skills
2. For businesses and organizations:
    a. to explore the potential benefits and limitations of Industry 4.0 solutions in their operations;
    b. to meet a group of informed academics that would support the development for new applications or improvements, and
    c. to receive support from other partners (i.e. businesses and researchers) in developing Industry 4.0 applications and identifying sources for funding their implementations.

Important aspects to consider were also food, technical support, prizes and spaces for students. These are reviewed in more detail in the following section.

Regarding ethical issues, no vulnerable participants were involved in this research, and no discussion on sensitive/personal matters took place. Researchers ensured anonymity by not collecting identifying information of participants (age, name, etc.); anyway, participation was voluntary, participants were free to withdraw from the study. Both paper and electronic data documents were secured/locked to ensure confidentiality, through OneNote for electronic data documents, and the university cloud (OneDrive) for a back-up copy. In both locations, documents were password protected. Additional security involved paper data documents locked in a filing cabinet inside the project coordinator desk. Participants provided an informed consent to communicate the conditions of their participation. Consent was also be obtained from the businesses that participated in this research with their challenges.

To Monitor and Control Group Work – Food and Technical Support

One of the main difficulties to monitor and control group work is to avoid interfering with the natural evolution of relations within the group. In the UoL4.0 Challenge 2019, the coordination team decided to make use of activities that would support students' work. These were providing food and technical support.

Students received free breakfast and lunch in organized events. This involved the main expenditure of the event, representing 47% of the budget. Important aspects were to find a provider able to deliver attractive menus for students; with flexibility to changes; a simple procedure for organizing what to have, when and where, and guaranteeing that everything would be on time and at good standards.

Students reacted very positively to the availability of food, as it gave them more time and a forum to focus on the challenge and, in some cases, better food quality than the one they usually have access to. The attendance to morning sessions with breakfast, and midday sessions with lunch, always had above 90% participant attendance.

Morning sessions also considered technical presentations. These were about (a) 'Inspired ideas', (b) 'Industry 4.0 revolution', (c) 'Professional abilities in a digital economy', and (d) 'How to present technical ideas/concepts to a business audience'. One lesson from this exercise was students' interest in attending technical presentations and making use of experts' knowledge.

In conclusion, to succeed in a CBL exercise the organizers need to consider how to maintain the contact with students, without interfering their work. One way is to provide scheduled collective activities where (a) additional information is provided and (b) additional resources that facilitate the progression of the work are made available to individuals, for instance food.

Other Issues – Spaces for Group Work and Prizes

From the start, it was clear that students required spaces for meeting and working together. This was confirmed as students made use of spaces, even though these were different from the ones negotiated for them, initially by the organisers. The coordination team spent almost a year in identifying whom to ask for spaces. In the particular case of UoL, the Library manages the learning lounges, small areas designed on purpose for group work, which are very attractive for students and always busy to use. However, the reservations made did not work as expected; as there was poor sign-posting and other students took the places. Anyway, students were very creative in finding alternative spaces, such as the theatre and the UoL interactive media lab. Another lessons was that if we need to properly support students with an effective working space, there is the need for funding associated to this resource as this would give us more control.

Another aspect we considered from the beginning was the idea of providing prizes. Even though the idea of a competition within the challenge was not fully shared, the coordination team agreed on providing awards and conducting an award ceremony at the end. Prizes represented the second biggest amount in expenditure (26% of budget). In this case, it was due to the small number of students and teams participating in the UoL4.0 Challenge 2019. The impact of prizes might be reduced proportionally depending on the quantity of students enrolled in the CBL event.

Anyway, it is important to note that students were pleasantly surprised by the prizes. In summary, we need to continue funding prizes, the most effective way seems to be inviting external organizations to sponsor them; for instance, in the UoL4.0 Challenge 2019, part of the prizes was funded by the Chartered Institute of Logistics and Transport (CILT).

Assessing the UoL4.0 Challenge

To assess the impact of challenge-based learning (UoL4.0 Challenge 2019) on students' learning experiences. Researchers conducted a thematic analysis of students' reflective accounts. A questionnaire with open-ended questions helped to understand what students thought about challenge-based learning in terms of its advantages and disadvantages. Table 3 shows the main issues indicated by the students.

Table 3. What should stay and what should change in future events

Q1. Best features of UoL4.0 Challenge 2019 as a learning experience	Q2. The shortcomings
Opportunity to: • Practice skills, such as teamwork, researching, communication, presentation and collaboration with cross-disciplinary students. • Work with other individual from different academic years and courses. • Gain some experience in presenting a proposal or pitch professionally • Know what is the real business working in the UK, and what is different between two different countries. I have learned a lot of experience during this challenge (international student). Good points: • It didn't take up much time, so it wasn't a large commitment out of other university work but you still managed to learn a lot and gain a lot of experience. • Morning breakfast was a great idea. • Linked to my operations management module, which kept me motivated throughout. • The whole experience allowed a lot of freedom, though timetables were set the Academics understood we have other priorities.	Teamwork: • More clear guidance on the roles of such people could be given. • Limited commitment from some of the participants Extra support: • More lectures, because they were very interesting and great to have right before meeting our team and talking about our ideas. • Could the tutor tell the student some ideas about each challenge during they are working with that, because I probably that much useful for students to open mind thinking about some new ideas. • Could be improved if the process was more long term and students were provided a budget with which they could try to implement their ideas for the chosen businesses or even just demo them. The budget money could be generated in many different ways; the Uni could arrange for temporary jobs for the students to raise the money or help organise other fund raiser type events. • Extra level of difficulty to the challenge to learn a new presentation style (Pecha Kucha). Organisation: • Better promotion! • Maybe it could be done a bit earlier in the academic year

Students enjoyed participating in the UoL4.0 Challenge and recognized several characteristics of this approach. First, its multidisciplinary nature: "the group really worked well [with] all different students from all different disciplines; we had interactive design students, we had a computer science student, we had a PhD student in bio med, and also a business management student". Second, UoL4.0 Challenge 2019 involved an open ended problem: "In my opinion, it was a nice, it was a focus problem that tells exactly what to do, but it was left open ended, which was nice, where we could think of what to do with it. Yeah, especially like in a team you do independent work and kind of bounce off each other's ideas and like, you know, done, what we kind of needed to". Third, it was technology-based: "... we were also able to use online, you know, software's like OneNote, you know, we also use Facebook messenger to communicate and yet they all came together really well at the end. It was definitely a good challenge". Fourth, UoL4.0 Challenge provided an opportunity to acquire some business acumen, for students that usually do not receive it in their education: "... nice kind of like introduction into how to pitch an idea to business and a nice way of like meeting new people and interacting with the businesses". Finally, CBL showed the importance of teamwork: "... it was mostly the team building exercises and be part been able to work with other courses".

Concerning the companies involved, there were messages of support and optimism in terms of what was achieved through UoL4.0 Challenge 2019. One of the managers commented that "... from our point of view, the project has been brilliant, we felt very lucky with a superb group of students who worked really hard on the project, spent quite a lot of time in the building and asked lots of very relevant and interesting questions". Another company manager indicated that "to see what the students have come up with in such a short time is quite phenomenal. We met one day, [...] then just about a week later, I'm seeing the solution, if you like, of their endeavors, and they've given us a really good idea that I want to explore further with them".

Accordingly, the success of the UoL4.0 Challenge 2019 was evident. As a consequence, CBL seems to begin to permeate in UoL educational practices. Other teams within UoL have organized two additional events, one focused on MSc students and the other on Lincolnshire-based students, in association with other educational and governmental institutions within Lincoln City. Furthermore, European funding has been achieved for continuing the UoL4.0 Challenge from 2020 to 2022 by the original organizers; this is discussed in more in detail in the next section. Moreover, some participating students have been successful in procuring the employment opportunities they applied to. Lastly, academics, themselves, learnt a lot, through cross-fertilization of ideas about new technology and multi teamed stakeholders.

FUTURE RESEARCH DIRECTIONS

Probably the most interesting outcome from the use of CBL at the University of Lincoln is its connection with long-term research. Research supported by 'Student as Producer' at UoL makes the research community wider and its impact deeper in the communities associated to UoL. As Morales-Menéndez et al. (2019) propose, long-term research projects can be associated with different actors from within universities (see figure 1).

Figure 1. Actors' potential contributions in long-term research projects within universities
Source: Adapted from Morales-Menendez et al. 2019

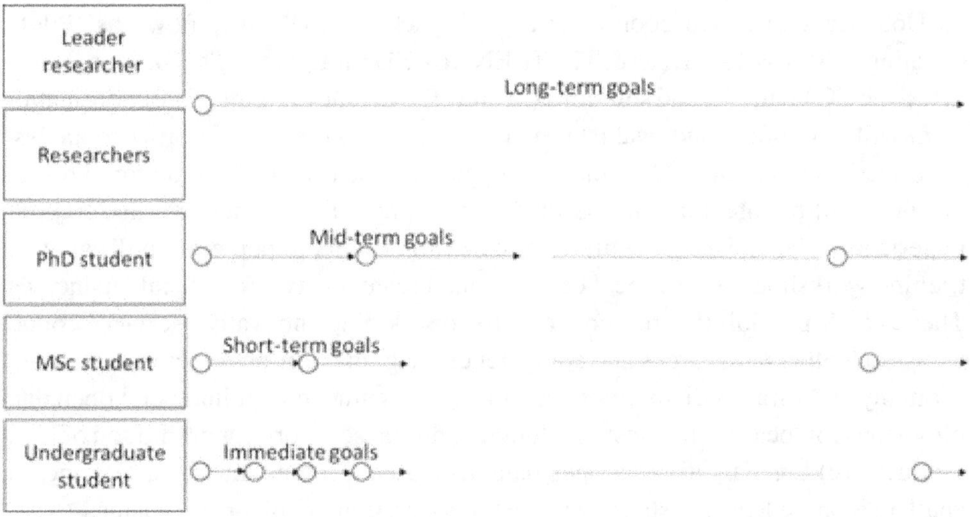

In the particular case of UoL4.0 Challenge 2019, one of the innovations was to consider students from all the different levels, BSc, MA, MSc and PhD.

The UoL4.0 Challenge 2019 made students central to discussions about technology non-adoption, to increase their exposure to real-world problems and improving their problem-solving skills. This event facilitated close working relationships between the University and SMEs in the region. It showed potential trajectories for improving the potential for adoption of digital technology and implementation of Industry 4.0 solutions among local SMEs, thereby improving business competitiveness and productivity.

The main outcome of the success running the UoL4.0 Challenge 2019 has been the collective interest on continuing with this project. Accordingly, the academic team decided to participate in an international collective bid for funding three future editions of the CBL and to disseminate the lessons learned. The project is known as COM^3, Building COMpetences for COMpet-itive COMpanies. COM^3 is a project, supported by the European Regional Development Fund (ERDF) that focuses on the mediatory role of local & regional authorities to improve tech-readiness and exploit local potentials for rural innovation and smarter growth. The COM^3 follows a training & coaching model to create 9 tech-enabled and tech-branded rural areas in the North Sea Region (NSR). COM^3 involves 19 partners that will test the jointly developed solutions in 9 regions from 7 NSR countries: Belgium, Denmark, Germany, Netherlands, Norway, Sweden and UK. The aim is to create an NSR roadmap and platform for digitalization of rural businesses, and build a cross-border community around it. The expected outcome is to increase tech adoption for rural businesses in a 15%, with a positive impact of ICT in the local/regional GDPs.

UoL academics will coordinate a work package 4 (WP4): known as 'Pilots: Creating TECH-BRANDED & TECH-ENABLED rural areas'. The focus in WP4 is to test the COM^3 model and solutions (from WP3) in nine (9) pilots. The coordination team will help pilots and evaluate their activities in all phases in order to analyse the effectiveness of the COM^3 model and guiding measures. The monitored results of pilots will be integrated in the model including guiding measures and training materials (WP3) and form the pilot stories. Furthermore, pilot regions will carry out training workshops series, local contests and create a network of local businesses. They search possibilities to cooperate for developing innovative service/product using each other's resources and competencies e.g. in traditional sector, data driven economy and rural tourism. Pilots include: (a) Drenthe, data mining and open data provisions for local businesses; (b) Hüttener, digital solutions for promotion of local products; (c) Leiedal, utilizing open data in creative/design industries; (d) Lincoln, challenge-based learning strategy in close cooperation of university, businesses and government from 2020 to 2022; (e) Oldambt, minimizing admin burdens for businesses by user friendly online service platform; (f) Torsby, digital tools and services for better visitors' experience; (g) Stavanger, digitization process optimizations in Asset Management; (h) Vejle, utilization of digital tools in traditional sectors based on circular economy principles, and (i) Vinje, increasing collaboration for business digitalization and reach new markets.

UoL4.0 Challenges 2020-22 will focus also on a series of events. A fundamental change will be that these events will take place over a semester. The coordination team do not foresee any problems with this change as previous experiences show their effectiveness in weekly and semester formats (Membrillo-Hernandez et al., 2018). At the beginning of the semester, the academic team will bring together SMEs, students and academics for initial discussions about the challenges. During consecutive weeks, students will work with SMEs in developing solutions to the challenges posed. Alongside these discussions there will be a series of workshops, delivered by University academics and external experts, to cover key themes regarding high-tech ideas development and effective engagement businesses in the digital industrial revolution. A final ceremony for presenting the solutions will take place on the final week. This ceremony will be open to the wider business community, as an opportunity for networking with other business peers, academics and other students.

The project will be led by the University of Lincoln's International Business School. It will be managed by a project coordinator and a core team of academics who will facilitate the series of Challenge Events for SMEs and undergraduate students at the University. The project will identify and recruit SMEs from the region to work on collaborative projects to identify Industry 4.0 solutions, ensuring a broad representation of size, sector and geographical location. The three groups of actors involved therefore are:

- SMEs, who supported by students' teams, will explore how connect their new commercial ideas with their potential markets; this by means of developing and using of digital technologies and applications.
- Students who will be able to increase their employability levels, by means of developing technology-based entrepreneurial and problem-solving skills, to better deal with challenges posed by cyber-physical systems.
- Academics who will be able to develop and test Challenge-based principles and examples that may be disseminated to other I4.0 communities around the world.

One expected effect of UoL4.0 Challenge 2020-22 is that the main beneficiaries from this pilot will be the same who will participate actively in it. In this sense, there is no expectation of external entities imposing particular views and agendas to the beneficiaries. The target groups will benefit from the pilots in the following ways:

1. For SMEs and organisations:
    a. to explore benefits and limitations to Industry 4.0 solutions in their operations;
    b. to receive state-of-the-art information on Industry 4.0 and related business projects;
    c. to reflect on the solutions proposed;
    d. to meet a group of informed academics that will support the development for new applications or improvements, and
    e. to receive support from other partners for identifying sources for funding such Industry 4.0 implementations.

2. For UoL students:
   a. to be part of the design of real Industry 4.0 solutions;
   b. to learn how to manage real technology-based business projects;
   c. to reflect on the solutions proposed, and
   d. to learn how to disseminate the lessons learned.
3. For Academics:
   a. To be able to develop and test new challenge- and project-based teaching approaches;
   b. To generate impact from existing research into Industry 4.0 solutions;
   c. To identify and disseminate good practice to teaching institutions to other institutions in the COM3 project, and to teaching institutions around the world.

However, it is noteworthy that there are also indirect beneficiaries from the pilot. Firstly, the UoL4.0 Challenge 2002-22 can be regarded as an exemplary application of the triple-helix model of innovation (Etzkowitz & Leydesdorff, 1995). In this context, the UoL4.0 Challenge may act as a laboratory for knowledge-based economic development, where relationships between university-industry-government will flourish and mature, and their effects be harvested. A practical instance of such positive effect is that the UoL4.0 Challenge looks at building on the concept of an 'entrepreneurial university'; one which "[...] takes a proactive stance in putting knowledge to use and in broadening the input into the creation of academic knowledge. Thus, it operates according to an interactive rather than a linear model of innovation" (Etzkowitz, 2003).

According to Elvekrok et al. (2018), expected outcomes from applying the triple-helix model of innovation are now limited when looking at participating firms. However, results from their study indicate that the primary benefits are increased access to knowledge and improved ability to meet challenges. However, they also indicate certain requirements such as having a committed manager and common activities that build relationships. Finally, Elvekrok et al. (2018) claim that firms' outcomes may be constrained by lack of resources; but participating in joint projects has a positive impact in their performance. This indicates collective efforts as UoL4.0 Challenge may benefit participating businesses.

CONCLUSION

Challenge-Based Learning is a teaching and learning approach that equips students with skills and competences for a digital working environment. The UoL4.0 Challenge 2019 successfully tested CBL principles to explore the impact of digital technologies in Lincolnshire economy. A proposal to connect digital ideas to the market was prepared. The aim of such proposal was to put Industry 4.0 opportunities within the reach of the UoL community, mainly students and businesses.

The project took two stages. The first one focused on investigating the possibilities of CBL as a teaching and learning approach appropriate for UoL practice, and developing a week-based challenge. The second stage involved the execution of the UoL4.0 Challenge 2019 and to test the principles of CBL. Students were organized in three teams and develop solutions concerning the implementation of digital technologies in different organizations. The outcomes of both stages were very positive and provided ample space to develop an international bid with 19 organizations located in seven North Sea Countries.

However, CBL is not free of risks and difficulties. Several lessons were learnt from this project. First, it is fundamental to understand that CBL is not a recipe that must be strictly followed. Flexibility is required in every aspect and involves testing issues such as format, schedule, and supportive guidelines, activities and materials. This conduce to a second aspect, the importance of including students and partners in the testing process. Different educational practices and cultural meanings directly affect how a CBL implementation must be designed and implemented. In the UoL4.0 Challenge, students were motivated to participate, even though they were not expecting impressive prizes. Providing a friendly environment, with additional guest lectures and food did a much better job in gaining students' commitment; however, this is not generalizable across different cultures and countries.

A point of attention concerns the answering who are the partners and were the work will take place. Partners need to be identified well in advance, with an interest linked to the challenge and an explicit definition of their level of commitment. A great organization committed to work with students, but unable to provide a challenge connected to the topic investigated, or the opposite, an organization full of challenging ideas, but with no commitment, should not be considered. This indicates the need to reflect on where students will be working together. An appropriate environment is required to maintain students' commitment and interest.

In summary, the UoL4.0 Challenge described in this paper concerns a series of elements, activities and decisions that may act as guidelines to support others in their quest to implement Challenge-Based Learning initiatives in their organizations. We sincerely hope this text will support their endeavor.

ACKNOWLEDGMENT

This research was supported internally at UoL by Lincoln Higher Education Research Awards 2018-19 and LIBS T&L Innovation Fund 2017-18 and 2018-19, and externally by the Chartered Institute of Logistics and Transport (CILT), the Industrial Digitalisation Project from HEFCE, Selenity and Siemens Industrial Turbomachinery Ltd.

REFERENCES

Apple. (2010). *Challenge-Based Learning. A classroom guide*. Retrieved from https://images.apple.com/education/docs/CBL_Classroom_Guide_Jan_2011.pdf

Asociación Española de Normalización, U. N. E. (2018). *Especificación UNE 0060:2018*. Retrieved from https://www.une.org/encuentra-tu-norma/busca-tu-norma/norma?c=N0060640

Dutra Moresi, E. A., Oliveira Braga Filho, M., Alves Barbosa, J., Carmo Lopes, M., Alves Tito de Morais, M. A., Alves dos Santos, J. C., . . . Osmala Júnior, W. A. (2017). O emprego do aprendizado baseado em desafios no desenvolvimento de aplicativos móveis [The use of Challenge Based Learning in mobile application development]. *2017 12th Iberian Conference on Information Systems and Technologies (CISTI)*, 1-6.

Elvekrok, I., Veflen, N., Nilso, E., & Gausdal, A. (2017). Firm innovation benefits from regional triple-helix networks. *Regional Studies*, *52*(9), 1214–1224. doi:10.1080/00343404.2017.1370086

Etzkowitz, H. (2003). Innovation in Innovation: The Triple Helix of University-Industry-Government Relations. *Social Sciences Information. Information Sur les Sciences Sociales*, *43*(3), 293–337. doi:10.1177/05390184030423002

Etzkowitz, H., & Leydesdorff, L. (1995). The Triple Helix -- University-Industry-Government Relations: A Laboratory for Knowledge Based Economic Development. *EASST Review, 14*(1), 14-19. Retrieved from https://ssrn.com/abstract=2480085

European Commission. (2017). *Germany: Industrie 4.0. Digital transformation monitor*. Retrieved from https://ec.europa.eu/growth/tools-databases/dem/monitor/sites/default/files/DTM_Industrie%204.0.pdf

European Commission. (2018). *The Factories of the Future*. https://ec.europa.eu/digital-single-market/en/factories-future

Johnson, L. F., Smith, R. S., Smythe, J. T., & Varon, R. K. (2009). *Challenge-Based Learning: An approach for our time*. Austin, TX: The New Media Consortium.

Membrillo-Hernández, J., Ramírez-Cadena, M. de J., Caballero-Valdés, C., Ganem-Corvera, R., Bustamante-Bello, R., Ordóñez-Díaz, J. A. B., & Elizalde, H. (2018). Challenge-Based Learning: The case of sustainable development engineering at the Tecnológico de Monterrey, Mexico City campus. *International Journal of Engineering Pedagogy*, *8*(3), 137–144. doi:10.3991/ijep.v8i3.8007

Ministero dello Svilupo Economico. (2019). *Piano Nazionale Impresa 4.0*. Retrieved from https://www.mise.gov.it/index.php/it/industria40

Morales-Menendez, R., Cantú Ortiz, F., Galeano Ramirez, N., Fangmeyer, J., & Hernández de Menéndez, A. M. (2019). *2019 IEEE Global Engineering Education Conference (EDUCON)*. Dubai, UAE: IEEE.

ProMexico. (2017). *Crafting the Future. A roadmap for Industry 4.0 in Mexico*. Retrieved from https://clusterinstitute.com/Documentos/MRT_Industry_I40.pdf

Rainer-Granados, J.J. (2018). *Informe LIBS Teaching Innovation Fund 2017/18. The UoL 4.0 initiative*. Internal communication.

Ramirez-Mendoza, R. A., Cruz-Matus, L. A., Vazquez-Lepe, E., Rios, H., Cabeza-Azpiazu, L., Siller, H., & Ahuett-Garza, H. (2018). Towards a Disruptive Active Learning Engineering Education. *2018 IEEE Global Engineering Education Conference (EDUCON)*. 10.1109/EDUCON.2018.8363373

Ringstaff, C., & Yocam, K. (1996). *Integrating Technology into Classroom Instruction: An assessment of the impact of the ACOT teacher development center project*. Apple Classrooms of Tomorrow. ACOT report #22. Retrieved from https://www.apple.com/euro/pdfs/acotlibrary/rpt22.pdf

University of Lincoln. (2016). *Thinking Ahead 2016 -2021. University of Lincoln. Strategic Plan*. Retrieved from https://www.lincoln.ac.uk/home/media/responsive2017/abouttheuniversity/managementandstrategy/UOL,Strategic,Plan,(MAR,2016),V5Final.pdf.

University of Lincoln. (2019). *Student as Producer*. Retrieved from https://studentasproducer.lincoln.ac.uk/

World Bank. (2019). *The Changing Nature of Work*. Retrieved from http://documents.worldbank.org/curated/en/816281518818814423/pdf/2019-WDR-Report.pdf

ADDITIONAL READING

Martin, T., Rivale, S. D., & Diller, K. R. (2007). Comparison of Student Learning in Challenge-based and Traditional Instruction in Biomedical Engineering. *Annals of Biomedical Engineering*, *35*(8), 1312–1323. doi:10.100710439-007-9297-7 PMID:17393336

Martínez, M., & Crusat, X. (2017) Work in progress: The Innovation Journey. A Challenge-based Learning methodology that introduces innovation and entrepreneurship in engineering through competition and real-life challenges. *2017 IEEE Global Engineering Education Conference (EDUCON)*. 10.1109/ EDUCON.2017.7942821

KEY TERMS AND DEFINITIONS

Big Idea: A wide concept that encapsulates an interesting issue for students and is pertinent to their communities. It must be able to be explored from different perspectives. Big ideas are not just questions bounded under specific sets of parameters, such as problems that is expected to be solved. The main source for inspiring big ideas rests in the joints between students' interests and their community needs.

Challenge-Based Learning: According to Apple (2010) is an engaging multidisciplinary approach to teaching and learning that encourages students to leverage the technology they use in their daily lives to solve real-world problems.

Essential Question: It is the articulation of the big idea, where it becomes a task associated to solving a particular problematic environment.

Chapter 8
Remote Laboratories for Engineering Education:
Experience of a Brazillian Public University With Project VISIR+

Isabela Nardi Silva
Universidade Federal de Santa Catarina, Brazil

Josiel Pereira
Universidade Federal de Santa Catarina, Brazil

Juarez Silva
https://orcid.org/0000-0002-5604-0576
Universidade Federal de Santa Catarina, Brazil

Simone Bilessimo
https://orcid.org/0000-0002-3326-8703
Universidade Federal de Santa Catarina, Brazil

ABSTRACT

The VISIR+ project was an international collaboration project for the dissemination of the remote laboratory VISIR, a tool to support teaching the theory and practice of electrical and electronic circuits. The initiative was first disseminated in Europe, and Latin American countries such as Brazil followed. This chapter essentially aims to discuss the experience of the Federal University of Santa Catarina in Brazil with the VISIR+ project. Various approaches were used for the dissemination of the initiative, including free courses for high school students, teacher training, and the creation of a virtual environment to discuss and share lesson plans that used the remote laboratory VISIR on their plots. In conclusion, the experience was observed as excellent for the institution and there was no reason to put the project ideas aside. After participating in the project, it becomes a challenge to ensure its sustainability.

DOI: 10.4018/978-1-7998-2562-3.ch008

INTRODUCTION

The VISIR + project was an international collaboration project that took place from 2015 to 2018. The initiative started at the Blekinge Institute of Technology (Sweden), with the aim of spreading the use of the VISIR remote laboratory to Latin America. The laboratory, which aims to support the teaching of theory and practice of electrical and electronic circuits, was widespread throughout Europe, so teachers using the tool had a lot of experience to add to new teachers. From this assumption came the idea of spreading the use of VISIR in Latin America.

Thus, the VISIR + project emerged, meaning "+" both because of the dissemination of the tool and the fact that the project was sponsored by an Erasmus + program consortium.

Among the institutions selected for participation in the project were the European Polytechnic Institute of Porto (IPP) in Portugal; Blekinge Institute of Technology (BTH), Sweden; National University of Distance Education (UNED) and University of Deusto, Spain. Among the Latin American institutions were the Brazilian Federal University of Santa Catarina (UFSC), the Federal Institute of Santa Catarina (IFSC) and the Pontifical Catholic University of Rio de Janeiro (PUCRJ) and the Argentine National University of Rosario (UNR).

At the Federal University of Santa Catarina, work on VISIR + was carried out within the Remote Experimentation Laboratory (RExLab), a research group founded in 1997 to promote the use of educational technologies and digital inclusion. The group represented the use of the remote laboratory within the University.

This paper aims to present the results acquired by the Federal University of Santa Catarina in relation to the interorganizational partnership provided by VISIR + project.

USE OF EMERGING TECHNOLOGIES IN EDUCATION

Information and Communication Technologies (ICT) are increasingly integrated into society's daily life. Today, it is hard to imagine how society would function without tools and features such as smartphones, tablets, desktops, laptops, among many others.

These resources act as facilitators of daily tasks, being teaching-learning just one of several areas that can be covered by the use of digital technologies.

Howard and Thompson (2015) define technology as all kinds of ICT capabilities in laptops, educational games or smartphones. These resources, which are increasingly present in the various actions taken in our daily lives, from access to news or communication with another person over a long distance.

From the integration of technology in the most varied areas, the incentive to integrate it also into education is understandable, since technology allows students to have control of their own learning (Montrieux et al, 2013).

In Brazil, the interest in including ICT tools in education arose in the 1960s from the Special Secretariat of Informatics, controlled by colonels linked to the National Information Service, whose function was to foster technological development in governmental sectors related to informatics and analyze and decide on projects to be developed related to computer goods (Brazil, 2017).

Later, other actions were developed, such as the National Program for Informatics in Education (PROINFO), in force between 1997 and 2006 (Silva et al, 2017).

The objective of this program was to promote the educational use of informatics in public elementary schools, bringing computers and educational resources to several Brazilian municipalities, where the municipal government would be responsible for providing the necessary physical structure and teacher education. According to data from the Ministry of Education of the Federal Government, the PROINFO program totaled 147,355 microcomputers acquired and 5,564 served over the 9 years of validity (Brazil, 2006).

More recently in 2010 the program “One Computer per Student” or “PROUCA” (SILVA ET AL, 2017a) was created. This program aimed to promote the inclusion of technology in federal, state, municipal, district and non-profit public schools, through the acquisition and use of technological solutions, made up of both hardware and software (Brazil, 2017).

However, despite the large investment made by government programs in the purchase of equipment for computer labs, it was found that these actions did not cover all municipalities in Brazil (Silva et al, 2017). In addition, due to the lack of maintenance of this equipment, over the years the laboratories become scrapped, making their didactic use impossible (Silva et al, 2017).

Data from the National Institute for Educational Studies and Research Anísio Teixeira (INEP) from 2015 state this fact, as they indicate that 45% of Brazilian public schools have computer labs, with an average of 7.3 computers per school (Silva et al, 2018).

On the other hand, draft laws were developed and laws were passed that prohibited the use of mobile devices in schools (Silva et al, 2018). According to Law No. 14,363, in force in the state of Santa CATARINA (Brazil, 2017), the use of mobile phones in the classrooms of public and private schools in Santa Catarina is prohibited. If technology has so many benefits in the daily lives of individuals, it is difficult to understand why such laws are passed (Silva et al, 2017). New technology has much to offer in basic education through the various resources that offer, and avoid them is bad for teaching development (Silva, 2016).

Faced with information and adverse practices to be daily, students may feel uncomfortable, something that culminates in the generation of difficulties, resulting in repetition and dropout, as stated by Neri (2009). Thus, it is necessary to search for solutions that motivate students, in order to present the content approached in the classroom in a more interesting way to them (Silva, 2016), using everyday tools to facilitate access to information and demonstrating how lesson content can be applied in real life (Silva et al, 2017).

Therefore, a good solution to consider is the use of technology in the classroom, because, besides using objects known in the students' daily life, it promotes a greater search for information and greater socialization of knowledge (Silva, 2016), providing the development differentiated classroom activities that broaden the concept of the classroom as well as the horizon of the students involved in the context.

REMOTE LABS FOR STEAM EDUCATION

Laboratory experiments bring science to life, representing the core of engineering learning in how they enable students to better understand scientific theories (Odeh et al, 2014). According to Corter et al (2011), laboratories can be divided into three categories: conventional laboratories, which have high associated costs, lacking space, technical support and infrastructure (Odeh et al, 2014); virtual laboratories, which reproduce real experiments with low precision, since the external environment does not influence experimentation (Silva et al, 2018); and the remote labs.

A remote laboratory is a type of experimentation in which the experimental apparatus and the user are physically separated, and the execution of the experiment depends on a means of communication (Internet) between the user and the remote laboratory, usually through a user interface (Fidalgo et al, 2014). Therefore, a student accessing a remote laboratory will have a very similar experience to accessing a traditional laboratory, as the result of his experimentation will vary depending on the current conditions of the environment where it is applied.

Experimental activities constitute one of the pillars for qualification at different levels of education, but their presence is essential in the scientific-technological and engineering areas (Silva et al, 2017). These activities help to improve understanding and enable mastery of scientific concepts and theories.

However, in many educational institutions, the lack of infrastructure for practical sessions significantly reduces experimentation (Roque, 2017), and this negatively impacts the quality of teaching, especially in the fields of science, technology, engineering, the arts. and math (STEAM).

Remote and virtual labs leverage wireless networks, mobile devices, and cloud-based software to make science experiments more accessible to schools that do not have fully equipped labs. In many ways, virtual and remote labs have benefits that hands-on labs do not have. For example, in virtual and remote environments, an experiment can be performed over and over again (Silva et al, 2015), with guaranteed access anytime and with more room to make mistakes, students can spend more time making scientific measurements and engaging. in laboratory practices (Silva et al, 2015).

Some emerging virtual and remote lab platforms incorporate report templates that inform experiment results so students and faculty can easily review results (Pereira et al, 2017). Educational institutions that do not have access to laboratory equipment can perform experiments and perform work online, accessing tools from any location (Lavechia et al, 2018).

In these labs, users are able to manipulate equipment or devices and watch activities unfold through a webcam on a computer or mobile device (Silva et al, 2015). This opportunity gives students a realistic view of system behavior and gives them access to professional laboratory tools whenever they need them.

Thus, it can be stated with conviction that virtual and remote laboratories present themselves as technological and pedagogical tools applicable at all levels of education (Pereira et al, 2018). That is, they have great potential for integrating technology in teaching and learning processes, due to their availability and also the affordable cost they present. The RELLE platform, developed by RExLab, is one of the platforms hosting the remote VISIR lab.

THE PLATFORM VISIR OPEN LAB

The Open Lab Virtual Instrument Systems (VISIR) Platform provides the functionality of a conventional electrical circuit lab that can be made available for remote access. This platform allows you to extend lab access to students inside and outside the IES 24/7 without any risk to the student or laboratory equipment. VISIR Open Lab is a client / server application that allows students to perform real electronics experiments using just a web browser.

The control / monitoring of physical equipment, connected to the server, is performed through virtual front panels and a virtual protoboard, both viewed through the client's computer. The server consists of an online workbench that provides multiple user access simultaneously by sharing access time to resources, giving users the impression that they are working with a real lab. This online workbench can complement a real lab equipped with oscilloscopes, function generators, digital multimeters, DC voltage sources, and protoboards.

In addition to the Blekinge Institute of Technology in Sweden, where the platform was developed other HEIs have already implemented the platform and are using it in their regular courses are: University of Deusto, Higher Institute of Engineering, Porto Polytechnic Institute, National University of Education Distance from Madrid, Carinthia University of Applied Sciences and Fachhochschule Technikum Wien (FH) in Austria, Indian Institute of Madras Technologies in India, College of the North Atlantic Qatar, among others.

Although the online workbench is being used predominantly by universities, it is also ideal for schools and vocational courses because of its flexibility, as it allows teachers to easily insert their own exercises. A dial-up modem connection and a web browser with Flash player are sufficient for the student to perform the experiments.

The software required to install the platform was published approximately one year ago under the GNU GLP license. In addition to a conventional PC, the hardware required to join the VISIR Project is to implement an online workbench using a PCI EXtensions for Instrumentation (PXI) chassis with National Instruments instruments and a switching matrix. The components that will be used by the students in the experiments are provided by the teachers and placed in the matrix.

The VISIR Platform can make a significant contribution to educating engineers with solid, well-documented laboratory experience, without significantly increasing the cost of training per student.

The Virtual Instrument Systems in Reality (VISIR) platform is a remote laboratory for teaching theory and practice of electrical and electronic circuits (Pereira et al, 2018). Therefore, it allows teachers and students to practice real experiments remotely and in real time, with equipment for testing and measurement, being accessed by the user through an interface (Fidalgo et al, 2014).

According to Diaz et al (2013), the biggest advantage of VISIR is that when compared to traditional electronic laboratories, it never has geographical or temporal restrictions.

Evangelista et al (2017) justify the application of VISIR in basic education because it is considered a tool of great potential to increase students' interest in certain particular topics, highlighting the fact that students are digital natives. Thus, the tool is focused on helping the teaching and learning of electrical and electronic circuits, while seeking to obtain added value in learning (Evangelista et al, 2017).

Therefore, it is clear that, although initially focusing on support for engineering education, VISIR can be applied to other contexts, such as applicability to other related undergraduate courses, as well as to high school students in the physics discipline. (Evangelista et al, 2017).

It is important to emphasize the importance of using VISIR in primary education as a tool of great potential to arouse students' interest in certain particular topics

(Carrara et al, 2018), in this case, the engineering area, besides adding to the student's learning.

Thus, it can be seen that the VISIR remote laboratory is a tool of great potential for application in physics education, for high school students as well as students of technical courses or undergraduate studies in STEAM areas.

VISIR+ PROJECT

From this scope emerges the VISIR + project, which was a consortium supported by the European Commission under contract 561735-EPP-1-2015-1-EN-EPPKA2-CBHE-JP, in the Erasmus + program area (Roque, 2017). The project aimed to define, develop and evaluate a set of educational modules that make up practical, remote and virtual laboratories.

This project was funded by the Erasmus+ program of the European Union and is being implemented in Brazil, Sweden, Spain, Argentina, Portugal and Austria.

The concept behind this project is that European institutions that used the remote laboratory (Blekinge Institute of Technology, Porto Polytechnic Institute, University of Deusto, National University of Distance Education (UNED) and Carinthian University of Applied Sciences) transfer their experiences at university-level Latin American institutions and associated institutions at the basic and higher levels (Branco et al, 2017).

In order to increase the scalability of the remote laboratory, some of the Latin American institutions have developed Didactic Practice Repositories, which house practices previously implemented in VISIR, such as the implementation described in Pereira's work (2018).

According to Pereira (2018), the repository provides remote lab configurations, tutorials and simulations of available practices. Being applied in six countries around the world, the project

VISIR + had 12 participating institutions. The Polytechnic Institute of Porto (Portugal), the University of Deusto and the National University of Distance Education (Spain), the Blekinge Institute of Technology (Sweden), the Carinthian University of Applied Sciences (Austria), the Federal University of Santa Catarina, the Federal Institute of Santa Catarina, the Brazilian Association of Engineering Education and the Pontifical Catholic University of Rio de Janeiro (Brazil) and the National University of Rosario, the National University of Santiago del Estero and the Rosario Institute for Science Research of Education (Argentina).

Thus, it can be seen that the VISIR + project consisted in exploiting the resources of education supported by remote laboratories, sharing the experiences of European groups with their Latin American partners (Arguedas-Matarrita et al,

2017), something that greatly enriches education. in engineering, facilitating the learning of students already enrolled and motivating more individuals to enter and careers in the STEAM areas.

INSTALATION OF VISIR REMOTE LABORATORY

In December 2016, the VISIR module was installed in the RExLab physical space, located at UFSC Campus Araranguá, as shown in the following figure (Pereira et al, 2017):

Figure 1.VISIR installed at REXLAB
Source: PEREIRA ET AL (2017)

Following the installation of the equipment, RExLab team technicians had the support and training of the Blekinge Institute of Technology regarding the operation and maintenance of VISIR.

The VISIR interface has been added and made available in RExLab through the Remote Labs Learning Environment (RELLE) platform available at http://relle.

ufsc.br/. This platform is responsible for managing users and experiments, and its functions include creating and editing remote users and experiments, reporting and access control to the platform (Simão, 2018). The figure below shows the RELLE platform home screen:

Figure 2. Access to VISIR via RELLE via a browser
Source: Authors

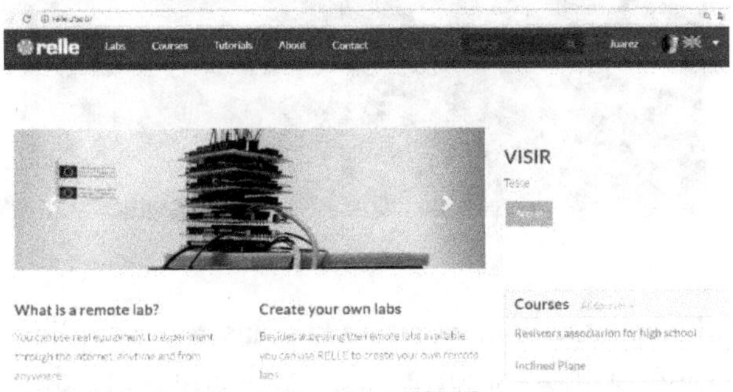

Access to the RELLE platform is open to users wishing to access it and to date has 18 remote labs available for access. From 09/2017 to 05/2018, RELLE's remote laboratories were accessed by 21,766 users. The accesses came from 1,431 municipalities in 121 countries. The system was developed in PHP 5.5 using the Model-View-Controller (MVC) Laravel framework on its back end, and its front end was developed in HTML, making use of CSS Bootstrap framework, in conjunction with the JavaScript Jquery library. (Simão, 2018).

The VISIR + project had some specific work packages, in which three distinct moments for training actions were established, namely Training Actions 1, 2 and 3 (TA1, TA2 and TA3) (Roque, 2017).

The first training action, TA1, was one of the initial milestones regarding the sharing of knowledge within the VISIR + project, as it allowed the first contact of Latin American partners with the VISIR remote laboratory.

This workshop was held at BTH, Sweden, in February 2016. For this event, all presentations and content were prepared and under the responsibility of all European partners, which characterizes a collaborative action between these institutions.

In TA1, technical, pedagogical and research aspects related to VISIR were addressed:

- **Technical Aspects:** Issues of operation and operational requirements of VISIR;

Figure 3. Team at TA1
Source: Authors

- **Pedagogical Aspects:** Content related to classroom applications, advantages of using VISIR, such as student-centered learning, student personal safety, among other significant points.
- **Research Aspects:** Procedures to be followed in the VISIR + project, such as student and teacher documentation and usage records.

At the end of TA1 participants were able to understand what VISIR is, how to use it, what experiments and practices it supports, how it can be incorporated into curricula of disciplines focused on electrical and electronic circuits, as well as the main learning outcomes. already existing.

Thus, regarding the RExLab team, TA1 was essential to obtain valuable information about VISIR and transform it into knowledge, as highlighted by Chiavenato (2009), and knowledge is information that can be understood by being Human because knowledge is in the minds of people who transform information into knowledge, whether through analysis and comparisons, making connections and communicating with others about the information they receive.

As well as the first training action, TA2 sought to share knowledge within the VISIR + project, but comprehensively as to the number of participating teachers, as well as the depth and detail of the contents taught.

In this training action were obtained professors from UFSC Araranguá, and from associates SATC and Instituto Federal Catarinense (IFC) Campus Sombrio. It was a training that enabled theoretical and practical information regarding VISIR. TA2 held at UFSC, Araranguá campus, in August 2016, was led by the RExLab team, which became responsible for the entire organization and organization of the event.

Figure 4. Team at TA2
Source: Authors

The teachers of IPP-ISEP, tutor of RExLab, actively participated in the training, in which they gave lectures and practical workshops using the VISIR remote laboratory.

The schedule was divided into two days:

1. **First Day:** The training was supported by the contextualization of the remote laboratory VISIR, which described its history and demonstrated its technical functioning. Therefore, for the public with technical and specific knowledge in electrical and electronic circuits, training was provided with the actual use of VISIR, enabling teachers to make real contact with VISIR through the elaboration of some practical examples.
2. **Second Day:** The training was marked by the presentation of examples of didactic implementations carried out by the European academic community, indicated advantages and challenges already known regarding the VISIR educational resource.

Participants received a guide providing a brief description of the actions required to prepare and conduct experiments involving electricity and electronics using the remote laboratory called VISIR.

The material was divided into three sections:

- **Section 1:** Introduction of the VISIR system, with its interface presentation, as well as the important procedures for preparing, setting up and conducting experiments;

- **Section 2:** Presentation of two practical examples of experiments that can be performed using the VISIR system, thus, it was shown how the students can act in the system, from the circuit assembly, measurements, and observations of the behavior of the electrical elements and electronics;
- **Section 3:** Suggestion for training participants, in addition to the examples shown, to propose and conduct new experiences in the VISIR system.

Therefore, the aim was to make it clear that VISIR enables students to conduct electrical and electronic experiments with the application of real equipment, which is found in a traditional laboratory, which allows the student to generate, measure and observe signals. in different situations problems associated with electrical and electronic circuits.

Each partner associated with the VISIR + project received the third and final training action, which focused primarily on knowledge sharing and the dissemination of the VISIR remote laboratory.

The RExLab research team organized TA3 to be held in three meetings:

- **Meeting 1:** It was held at the UFSC headquarters, Araranguá campus, and the teachers of the institution itself and the teachers of the IFC Campus Sombrio were invited.
- **Meeting 2:** Was held at the premises of the associate SATC, for the event were invited and attended higher education teachers of undergraduate courses in electrical engineering, mechatronics, mechanics, and technology in industrial automation, in addition to teachers of vocational education of technical courses in electrical engineering, electronics, mechanics and automotive.
- **Meeting 3:** Was aimed at the general public, in which the invitation was extended to several regional educational institutions, which have real potential for the use of VISIR in their courses.

In other lines, teachers from institutions capable of becoming future partners to this project participated. It occurred in the dependencies of UFSC itself.

Figure 5.Team at TA3
Source: Authors

The three meetings that constituted TA3 were similarly understood as to objectives, and thus had equivalent content and materials. Thus, all meetings were started with the theme about the importance of technology in engineering education, lecture given by the vice president of ABENGE, Mr. Luiz Paulo Brandão.

The Lecture on the importance of technology in engineering education focused on sharing knowledge about the use of technologies applied as educational resources and their importance, as well as alerting teachers about the constant changes in education, such as student profile, new teaching and learning methodologies, academic autonomy, among other current trends that need to be discussed in academia.

Thus, this part of the meetings aimed to contextualize and enable participants to recognize the VISIR laboratory, showing them when, how and who was responsible for its creation, which educational institutions are already using it, what are the advantages obtained with this feature, what are its main technical characteristics, and lastly, what is required to use it in your classes.

INTRODUCTION TO ROBOTICS COURSE

In order to stimulate young people's interest in science, technology, engineering, arts and mathematics (STEAM) disciplines and, consequently, in related professions, RExLab held the Robotics Introduction Workshops with students in classes. from the Maria Garcia Pessi Primary School, a public school located in the municipality of Araranguá / SC (Carrara et al, 2018).

The workshops were held in face-to-face meetings, where after a brief explanation of the concepts, students controlled a protoboard and various electronic components (Silva et al, 2017). In addition to classroom practice, students were encouraged to access the remote VISIR lab in an out-of-class environment as a way to review the topics addressed (Carrara et al, 2018).

The Introduction to Robotics workshop started in June 2017 and enrolled approximately 40 first year high school students, of whom only 13 completed the workshop (Silva et al, 2017).

Classes were held in the school's computer lab during the counter shift. Thus, the material for practical and theoretical classes was moved from RExLab to school (Silva et al, 2017). The contents were divided and structured so that every subject was accompanied by one or more practical classes, promoting better learning, more efficient fixation of contents, greater interaction between participants, stimulation of scientific thinking and a series of other factors aimed at encouraging STEM areas (Carrara et al, 2018).

Figure 6. Students attending Robotics Introduction workshop
Source: Authors

Among the theoretical and practical classes, there were lists of exercises, which were solved by using some of the remote laboratories, both for help and for better exemplification and even in the demonstration of situations (Carrara et al, 2018). The accesses were essential for the accomplishment of the activities, since the experiments were the only alternative for the confection of the circuits, data extraction and aid in the calculations.

In practical classes, students received representations of electronic circuits and applied them to protoboard, with the help of the other participants and the lecturer (Silva et al, 2017). Upon completion, the students explored new combinations of components, new values of voltage, current, resistance, among others. Some practices were also aided by VISIR, such as the use of the oscilloscope and the function generator (Silva et al, 2017).

The following figure shows a photograph of one of the workshop classes:

At the end of the workshops, it was clear that the actions performed contributed to the learning of the students involved, as it provided greater exercise of logical reasoning and cognitive ability, and consequently, a greater educational development (Carrara et al, 2018).

VIRTUAL COMMUNITY OF PRACTICES: LABS4STEAM

The Labs 4 STEAM platform was developed using Dokuwiki, an open source wiki software containing a large number of plugins (Dokuwiki, 2017). According to Dokuwiki's official website homepage (2018), the software is admired by its users for its clean and easy-to-read syntax, as well as its easy maintenance, backup and integration. Built-in access control and authentication connectors make Dokuwiki especially useful in the corporate context and the high number of plugins built collaboratively by your community (Dokuwiki, 2018).

Initially, the platform would only provide access to lesson plans related to the use of the remote VISIR lab. However, in thinking about the platform's expansion and greater popularity among its future users, it was decided that the platform would also accept other remote labs, and in the future even simulations.

The platform was developed during the first half of 2018 and is available for access in September 2019 and is available at: http://labs4steam.rexlab.ufsc.br/ A difference in permissions has been established between registered users and visitors to strengthen the platform's security and to prevent possible interventions by an unselected audience to use it. Only teachers may discuss and submit new practices. The administrator checks each platform registration request.

The platform is available in English, Portuguese and Spanish. However, for purposes of limitation, not all practices are available in all languages. The platform allows the translation of its interface, but not the content posted manually by the user. Therefore, it was proposed to users that the platform is willing to receive volunteers who can translate the practices into other languages.

The platform registration form was developed using the Google Forms tool and consists of the following questions: full name, date of birth, email, training and subjects you teach. After completing the form, the applicant will receive their username and password by email within 24 hours. From the moment they are registered, the user is allowed to comment on the platform. If not registered, the user can still download lesson plans and research them, however, for security reasons it was determined that only registered users could make comments.

The form for submitting lesson plans to Labs4STEAM was also developed using the Google Forms tool. It includes: full name, education, e-mail, lesson plan course, other subjects you teach, and upload lesson plan to .PDF, .DOC or .PPT, and the like. It was defined that the plans would be sent via form, not entered directly by users to the platform, to control the material to be posted on the platform.

A section called "About Labs4STEAM" has also been inserted with information about Labs4STEAM to explain its origin. The section also features some platform outreach videos.

CONCLUSION

From this chapter, it can be seen that the results of RExLab's partnership with the VISIR + project was to provide VISIR dissemination.

Deploying VISIR on RExLab was paramount for project amplification as it increased one instance of the feature. In addition, it allowed for greater diversification of the content provided by the machine, allowing students and teachers to access it for several reasons.

With the training actions, it was possible to train teachers and researchers regarding the use of the remote laboratory, its functionalities and application. The three training editions provided a fuller understanding of both the VISIR tool and its design.

From the workshops on introduction to robotics, it was possible for young people to get to know the tool, as well as exemplifying the potential of VISIR as a resource to encourage students to enter STEM courses.

The elaboration of the Labs 4 STEAM: Collaborative Community of Practices online platform allows discussion of lesson plans that make use of both VISIR and other remote labs and even simulations. This brings good to many teachers who sometimes give up on innovating in the classroom because of the lack of dialogue with other teachers; It also allows new teachers who do not recognize the tools or are new to having a downloadable repository to base their lessons on. A tool like this is a great opportunity for those teachers that have so much to share, but sometimes don't have this opportunity in their daily lifes.

Therefore, it can be seen that RExLab's experience with VISIR + was to open doors for the sustainability of the VISIR + project and to allow greater dissemination of the VISIR educational tool. Sources like this are very important for education and it is fundamental to keep them alive so people can have more opportunities to teach and learn.

REFERENCES

Arguedas-Matarrita, C., Concari, S. B., García-Zubia, J., & Marchisio, S. T. (2017). *A teacher training workshop to promote the use of the VISIR Remote Laboratory for electrical circuits teaching*. Presented at the 4th Experiment@ International Conference (exp.at'17), Portugal. 10.1109/EXPAT.2017.7984351

Branco, M. V., & Coelho, L. A. (2017). *Aspectos de diferenciação entre laboratórios remotos e simuladores*. Presented at the Congresso Brasileiro de Educação em Engenharia - COBENGE, Brazil.

Brazil. (2017). *Lei n. 14.363 de 25 de janeiro de 2008*. Retrieved from http://www.leisestaduais.com.br/sc/lei-ordinaria-n-14363-2008-santa-catarina-dispoe-sobre-a-proibicao-do-uso-de-telefone-celular-nas-escolas-estaduais-do-estado-de-santa-catarina

Carrara, A. C., Silva, I. N., Lotthammer, K. S., Silva, J. B., & Ferenhof, H. (2018). *O uso do laboratório remoto VISIR em oficinas de eletrônica para estudantes do ensino médio*. Presented at the Congresso Brasileiro de Educação em Engenharia - COBENGE, Brazil.

Corter, J. E., Esche, S. K., Chassapis, C., Ma, J., & Nickerson, J. V. (2011). Process and learning outcomes from remotely-operated, simulated, and hands-on student laboratories. *Computers & Education*, *57*(3), 2055–2067. doi:10.1016/j.compedu.2011.04.009

Dokuwiki. (2019). *About Dokuwiki*. Retrieved from https://www.dokuwiki.org/pt-br:dokuwiki

Farina, J., Marchisio, S. T., Concari, S. B., Lerro, F. G., Pozzo, M. I., Alves, G. R. C., ... Nilsson, K. (2017). *Preparando estudiantes secundarios para carreras de ingenieria: un estudio de caso utilizando el laboratorio remoto VISIR*. Presented at the Congresso Brasileiro de Educação em Engenharia - COBENGE.

Fidalgo, A. V., Alves, G. R., Alves, M. A., & Viegas, M. C. (2014). *Adapting Remote Labs to Learning Scenarios: Case Studies Using VISIR and RemotElectLab*. IEEE Revista Iberoamericana de Tecnologias Del Aprendizaje.

Gustavsson, I., Nilsson, K., Zackrisson, J., Garcia-Zubia, J., Hernandez-Jayo, U., Nafalski, A., ... Hkansson, L. (2009). N objectives of instructional laboratories, individual assessment, and use of collaborative remote laboratories. *IEEE Transactions on Learning Technologies*, *2*(4), 263–274. doi:10.1109/TLT.2009.42

Gustavsson, I., Zackrisson, J., & Lundberg, J. (2014). *VISIR work in progress*. Presented at the *IEEE Global Engineering Education Conference (EDUCON)*, Turkey. 10.1109/EDUCON.2014.6826253

Lavechia, J., Silva, J. B., & Spanhol, F. J. (2017). *Publicações científicas do Laboratório de Experimentação Remota – RExLab: uma revisão sistemática*. Informação & Informação.

Lima, N., Viegas, C., Alves, G., & Garcia-Peñalvo, F. J. (2016). *VISIR usage as an educational resource*. Presented at the TEEM '16, Spain.

Lobo, M. C. C. (2011). *Using Remote Experimentation in a Large Undergraduate Course: Initial Findings*. Presented at the *41st Asee/ieee Frontiers In Education Conference*. 10.1109/FIE.2011.6142913

Nickerson, J. V., Corter, J. E., Esche, S. K., & Chassapis, C. (2007). A model for evaluating the effectiveness of remote engineering laboratories and simulations in education. *Computers & Education*, *49*(3), 708–725. doi:10.1016/j.compedu.2005.11.019

Pereira, J. (2018). *Implantação de módulos educacionais para circuitos elétricos e eletrônicos em universidades brasileiras no âmbito do projeto VISIR+*. Retrieved from https://repositorio.ufsc.br/handle/123456789/189331

Pereira, J., Silva, I. N., Simão, J. P., Carlos, L. M., Silva, J. B., & Bilessimo, S. M. S. (2017). *Modelo de repositório de práticas didáticas de circuitos elétricos e eletrônicos usando o laboratório remoto VISIR*. Presented at the Congresso Brasileiro de Educação em Engenharia - COBENGE.

Rodrigues, A. L. (2016). A integração pedagógica das tecnologias digitais na foração ativa de professores. Presented at the *IV Congresso Internacional das TIC na Educação: Tecnologias digitais e a Escola do Futuro*.

Roque, G. R., Silva, I. N., Alves, G. R., Alves, J. B., Silva, J. B., & Bilessimo, S. M. S. (2017). *Utilização do laboratório remoto VISIR como recurso educacional num curso de engenharia mecatrônica*. Presented at the Congresso Brasileiro de Educação em Engenharia - COBENGE.

Salaheddin, O. (2014). *Experiências da Aplicação de VISIR na Universidade de Al-Quds*. Presented at the *11th International Conference On Remote Engineering And Virtual Instrumentation*, Portugal.

Silva, I. N. (2016). *Uma proposta de aplicação de experimentação remota e dispositivos móveis em aulas experimentais de física no ensino médio*. Presented at the SICT-SUL.

Silva, I. N., Pereira, J., Bilessimo, S., Alves, G., & Silva, J. B. (2018). *O Laboratório VISIR: uma revisão sistemática*. Presented at the Simpósio Íbero-Americano de Tecnologias Educacionais.

Silva, I. N., Roque, G. R., Bilessimo, S. M. S., Alves, G. R., & Silva, J. B. (2018). *Laboratório Remoto VISIR: estudo de caso da integração na plataforma RELLE*. Retrieved from https://www.aacademica.org/congreso.aahd2018/tabs/abstracts

Silva, K. C. N. (2018). *Inovação social na educação básica: um estudo de caso sobre o Laboratório de Experimentação Remota da Universidade Federal de Santa Catarina*. Retrieved from https://repositorio.ufsc.br/handle/123456789/191134

Silva, K. C. N., Silva, I. N., & Bilessimo, S. M. S. (2017). *Oficinas de eletrônica básica para estudantes do ensino médio de uma escola pública de educação básica*. Presented at the Simpósio Íbero-Americano de Tecnologias Educacionais.

Silva, P. F., & Melo, S. D. G. (2018). O trabalho docente nos Institutos Federais no contexto de expansão da educação superior. *Educação e Pesquisa*, *44*(0). doi:10.15901678-4634201844177066

Silveira, M. S., & Carneiro, M. L. F. (2012). *Diretrizes para a Avaliação da Usabilidade de Objetos de Aprendizagem*. SBIE.

Simão, J. P. S. (2018). *Modelo para registro de dados de experiência de aprendizagem em laboratórios remotos*. Retrieved from https://repositorio.ufsc.br/handle/123456789/191118

Tawfik, M. (2013). Virtual instrument systems in reality (visir) for remote wiring and measurement of electronic circuits on breadboard. *IEEE Transactions on Learning Technologies*. 10.1109/TLT.2012.20

Wenger, E. (2010). *Communities of Practice and Social Learning Systems: the Career of a Concept*. Social Learning Systems And Communities Of Practice.

Chapter 9
How Serious Games Contribute to the Learning Experience of Engineering Students and Professionals:
A Serious Supply Chain Management Game

Patrick Willems
UC Leuven-Limburg, Belgium

ABSTRACT

Engineers in a globalizing world need innovative skills as well as learning capabilities. They often need to cooperate in different teams in the supply chain of a company. Thus, they should practice before they perform like athletes and musicians do, but they cannot practice by making mistakes since it would be an expensive way to learn. Serious business games serve as a platform for the exchange of ideas, the sharing of expertise, and the alignment of objectives. Higher productivity along with employee satisfaction are their most prominent outcomes. Engineers should be able to work in teams, where behaviors can have both positive and negative effects. Serious business games can support organizations to develop more effective team behavior, influence the level of cooperation in a group, and, in turn, increase the company's profit. In this study, the authors examine a supply chain serious game called "The Fresh Connection" and discuss how it can improve the learning process at the university and further allow people to continue their learning process as an engineering professional.

DOI: 10.4018/978-1-7998-2562-3.ch009

INTRODUCTION

Engineers are confronted with a growing complexity and an exponentially growing information flow. Weenk (2019) mentions in his book "Mastering the supply chain", that when you want to develop skills, experiential learning becomes more and more important. We live in the age of accelerations, the world is changing faster and faster, calling for other skills than the ones that were valid in the past (Friedman, 2016). The need for training 21st century skills was expressed on different occasions (World Economic Forum, 2016, Robinson and Aronica, 2015).

The World Economic Forum published their view on "21st-century skills" (WEF, 2018), including the following:

- Analytical thinking and innovation
- Active learning and learning strategies
- Creativity, originality and initiative
- Technology design and programming
- Critical thinking and analysis
- Complex problem-solving
- Leadership and social influence
- Emotional intelligence
- Reasoning, problem-solving and ideation
- Systems analysis and evaluation.

Most of these skills are related to complexity and how to deal with it, as well as human interaction. In order to deal with the increasing complexity in the world, we will need to work more in teams. Experiential learning is an appropriate way to train such skills (Kolb, 2014). Kolb's learning cycle is presented in Figure 1. The main idea behind the learning cycle is that "knowledge results from the combination of grasping and transforming experience. Grasping experience refers to the process of taking in information, and transforming experience is how individuals interpret and act on that information. [...] This process is portrayed as an idealized learning cycle or spiral where the learner 'touches all the bases'." (Kolb, 2014)

In experiential learning, the focus is on first-hand experience, which allows for reflection on what happened and why, leading to a conceptual view of the situation, potentially reinforced by existing theories and/or frameworks. This combination will then be the basis for an improved view on the situation, which can then be applied in the next experience, either in class or other study environment, or directly in a real-world situation.

Figure 1. The learning cycle
Source: Kolb (2014)

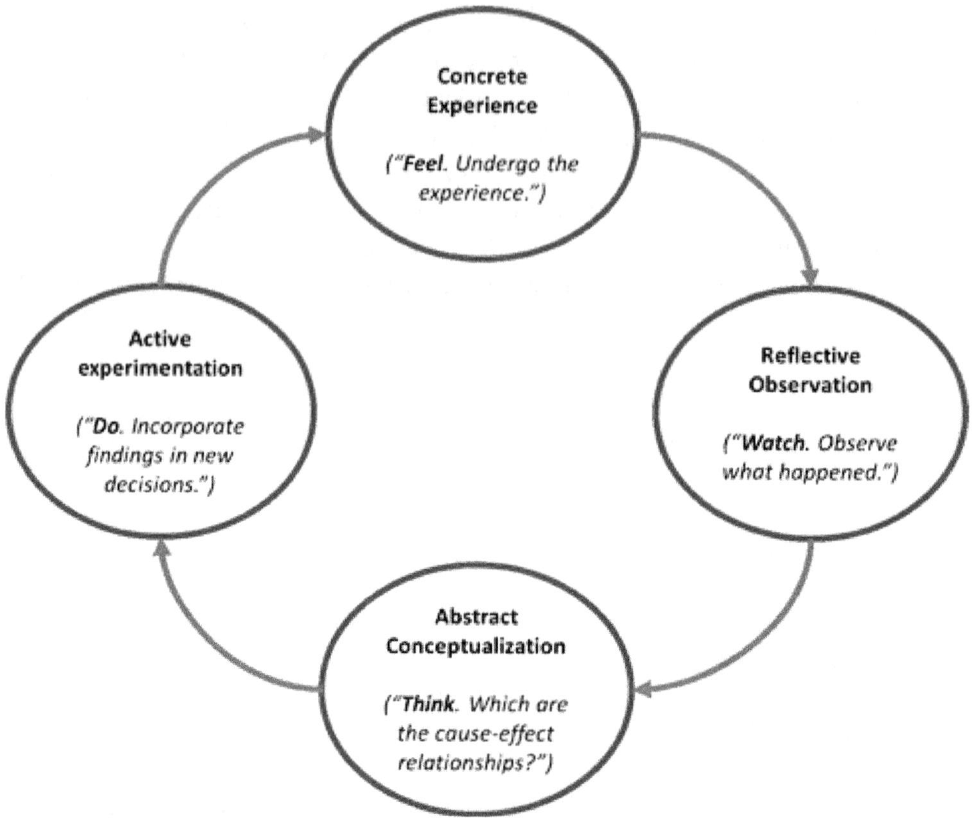

Experiential learning is also based on methodologies such as the case method, championed by Harvard Business School, on in-class teamwork as well as on serious gaming.

Serious gaming is playing games that are designed for a primary purpose other than pure entertainment. Serious games are a subgenre of serious storytelling, where storytelling is applied "outside the context of entertainment, where the narration progresses as a sequence of patterns impressive in quality ... and is part of a thoughtful progress" (Lugmayr et al., 2017). It explicitly emphasizes the added pedagogical value of fun and competition.

A supply chain is a complex system with many companies operating in a dynamic and uncertain environment (Serdarasan, 2013). Managing a supply chain considering the need to plan and execute a high variety of processes and interactions between and within the companies, where a huge amount of information required to control the whole system is a difficult job. Management simulation games are frequently used within companies and in business schools for development and learning purposes. They often offer a first experience in supply chain, such as the Beer Game or the Wood Supply game and the Fresh Connection, but even for experienced people, such a business game can still be a challenge. As the students and practitioners keep on learning about the complexity and sensibilities of supply chains, they realize that there is not only the technical logistics part of a supply chain, but also it is about collaboration and team management

The objective of this chapter is to present the influence of "The Fresh Connection" game on lifelong learning by individuals and groups, both in higher education and professional context. 'The Fresh Connection' (TFC) is a computer- based supply chain simulation game which can be classified as a serious game, which is a cognitive tool for learning and development, but playful and fun at the same time. (Hummel et al, 2011) TFC is a virtual fruit juices producer that is suffering severe losses. The Chief Executive Officer (CEO) of the company is searching a new management team to revive the company. He is willing to change his whole supply chain in order to obtain the highest possible Return On Investment (ROI).

The chapter discusses learning and cooperation in teams, how serious games facilitate the learning process and briefly describes the supply chain processes. Then it explains in detail how "The Fresh Connection" game works and present some learnings from different researches and experiences of players.

BACKGROUND

The background section focuses on three topics: learning in teams, serious business games, and supply chain management processes.

Learning and Cooperation in Teams

"If everyone is moving forward together, then success takes care of itself". Henry Ford

This quote by Henry Ford makes it clear that effective cooperation between individuals can create competitive advantage for organizations (Watson et al., 1998). The increasing competitive pressure and the fast technological progress make it extremely important but also difficult to gain this competitive advantage and to keep it (Allred et al., 2011).

Competition in the current environment has to go further. "The possession of means of production", as Allred et al. (2011) state, "is a necessary, but not a sufficient condition to gain competitive advantage". Therefore, companies are more and more busy with their transformation and optimization process with the effective cooperation of company divisions and functional departments (Hansen & Nohria, 2004). Hansen en Nohria (2004) acclaim this trend by stating that effective cooperation can be a source of competitive advantage, as it is something that is not automatically available in a company.

The hypothesis is then that with the help of dynamic cooperation one can gain this extra length (Schippers et al.; Allred et al., 2011). Where the dynamic aspect of this term aims at the possibility to adapt to constantly changing environments, and cooperation refers to the integration, alignment and improvement of internal and external competencies (Allred et al., 2011). These characteristics of dynamic cooperation is what one needs to reach production growth and development. They are also considered in the Sales & Operations Planning (S&OP) of a company. S&OP is a process that encompasses the planning of the supply chain operations of a company. An effective and dynamic cooperation between business functions is needed so that profit (and competitive advantage) can be gained.

The exchange of ideas and competencies leads to alignment within the cooperation, which leads to optimisation of performance, capacities and possibilities of the employees (Cotter et al., 2009). Because there is more exchange and discussion, employees will investigate things deeper and more critically (Ku et al., 2013).

One can thus state that making decisions in a team is essential for the performance and the competitive advantage of the team (Schippers et al., 2012). West (2012) states that a team which communicates and collaborates effectively, which exchanges ideas, distributes workload evenly and gives support, becomes a dream team that can reach its full potential. Available means and capacities can be put to work more effectively (Hansen & Nohria, 2004).

Not only the team but the whole organization can become stronger, thanks to the success of one team that can inspire the rest of the teams within an organization to collaborate in a better way and to reach similar successes (West, 2012).

Already in the eighties, one experienced that group performances are better on qualitative and quantitative level than average performances on individual level.

Johnson & Johnson (1994) state that groups learn faster, make less mistakes and better decisions through the ideas, insights and strategies that come out of group discussions (see Watson et al., 1998). Also group members push each other and help each other to perform better, while amongst individuals competitiveness and fear for punishment or failure leads to worse performances (Baron et al., 1992; see Watson et al., 1998). Hansen en Nohria (2004) synthesise this by stating that by means of cooperation one can achieve what one cannot achieve as an individual.

According to a review, Hill (1982), research has proven that groups need fewer different solutions before finding the right one. As an explanation one states that groups can put together all kinds of information from different sources, by which errors counterbalance each other or are more quickly improved/remarked. The disadvantage of making decisions in a group is that they need more time than an individual to come to a decision, even though the process to reach a decision becomes easier in a group (Hill, 1982).

Based on the literature, one can state that results – in this case the ROI (Return on Investment) – of individuals would be lower than the team result. Overall we find more advantages of cooperation in the literature than disadvantages.

When one joins a company, one has to cooperate with new people. Some groups that have played the game, did know each other beforehand. The question is if this influences cooperation in a significant way. Myaskovsky et al. (2005) show the importance of group training so we can state that knowing each other beforehand has a positive impact on cooperation. They prove this by stating that regular contact (frequency of contact by Festinger et al., 1950) can lead to team members discovering each other's strengths based on actual experiences instead of stereotypes and this can lead to cooperation in a more effective way. West (2012) points out that teams become stronger as they work together during a longer period. It gives them the possibility to grow and develop and to clearly define each role. So we can conclude that knowing each other beforehand can lead to better performance.

Also tacit knowledge can be dispersed, optimised and maintained through communication and cooperation. Organisations have come to appreciate more this knowledge, because it gives an added value in the solution of problems (Hummel et al., 2011).

Tacit knowledge is knowledge that is not documented, but still present in the company. It is a kind of abstract knowledge that is acquired unconsciously, but can lead to big advantages for a company. It can help solve problems and make the right decisions. For example, in situations where rapid reaction is required or in very complex situations because there is no time in these circumstances to look for documented knowledge (Reber, 1989).

The dispersion of (tacit) knowledge, ideas and competencies also is considered crucial within Sales & Operations Planning (S&OP), where different teams (sales, production, finance, …) come together to forecast demand, develop new products, select new suppliers and where quality control is discussed (Schippers et al., 2012; Lapide, 2005). They make decisions with regard to the supply chain management from raw materials to the delivery of finished products to the customer (Oliva & Watson, 2009). They are responsible for a good alignment between supply and demand, such that profit or competitive advantage can be gained (Schippers et al., 2012)

Originally one thought information technology would facilitate this planning process, but this was not the case (Lapide, 2005). Investments and efforts in bridging the gap between different departments were seen as unsuccessful. Managers did hardly see any progress, despite of the technological progress they made in function of better communication (Allred et al. 2011). It is therefore important that different departments within S&OP can collaborate with each other as a team and that they collaborate in such a way that the bad decisions can be distinguished from the good ones and that one comes to a plan where supply and demand are in balance (Schippers et al., 2012). Because research from Allred et al. (2011) shows that even the smallest changes made in function of a better cooperation, lead to an increase in productivity and satisfaction, the search for a better way to collaborate remains necessary.

Schippers et al. (2012) investigated this with the help of "The Fresh Connection Game" and concluded that good planning can be achieved by stimulating the team to think about objectives, their functioning and the used methodologies. He calls this 'reflexivity', which can be translated as reflection and can be defined as the "tendency to reflect critically about goals and strategies" (De Dreu & Beersma, 2010). Reflection can lead to discussion about bad decisions so that they can be identified and discarded (De Dreu & Beersma, 2010; Schippers et al., 2010). This would make teams more productive and effective, leading as well to more innovation (West, 2012).

According to research at RSM/Delft University of Technology ", team reflexivity is a team's ability to consciously and reflexively react to changing and fluid situations and adapt accordingly". This ability benefits teams whose mix of team members tends to favour seeking "accomplishment and attaining positive outcomes, and where individuals are more inclined to explore all possible means". This in contrast with teams where team members rather than on attaining positive outcomes tend to focus mainly on avoiding negative outcomes (Schippers et al, 2011).

Good cooperation can thus be seen as the feeding ground for the exchange of ideas (Cotter et al., 2009) so that good or better elaborated decisions can be made. (De Dreu & Beersma, 2010; Schippers et al., 2010) and – as a consequence of the increased exchange of information, expertise and means as well as the alignment of underlying goals – an increase in productivity and satisfaction can be reached thanks to lower costs, higher quality, improved customer service and satisfaction, as well as higher added value in relations (Allred et al., 2011). Also the creation of positive feelings within a group would lead to a more connected group. This would then not only lead to better decisions, but also to a situation in which one would help each other (in teams as well as in organizations); one will do more than one expects from one's own function (West, 2012).

As a future leader, the engineer wants to reach more than results. He wants to grow together with his people and become stronger. That's where you need to estimate the potential of your group and cannot put the target too high.

Effective leaders want to change their team to a learning group, whose members strive continuously to increase their competencies.

One should create a climate where the group members reflect and learn from experiences, both from successes and failures, and where team members support and help each other. Team members should be allowed to freely express their proposals and opinions and encouraged to develop innovative and creative solutions and ideas. Thus you increase motivation and involvement of your team members.

Mutual trust should be reinforced by personal attention, support and encouragement. You should know the needs, capacities and aspirations of your people and you try to develop their competencies through trainings, coaching or mentoring.

From MIT-research we learn that teams where the individual team members independently from one another indicate high levels of perceived trust create a good working atmosphere leading to better results. They also indicate "the fragile nature of trust" and, in the case of virtual working teams with individual team members being located in physically different places, "the sharp improvement after the team-member had met face-to-face", suggesting that team members that did not know each other previously and who until then had only communicated via email, telephone, videoconference or the like, started working together much more productively after actually having met in person. According to the research, meeting face to face gave an important boost to mutual inter-personal trust between the team members (Phadnis et al, 2013). Thus you'll form a team that can adapt flexibly to a changing environment.

Serious Games in Education and Training

For the generations now in school (i.e. Generations Y and Z), traditional education methods fail regularly. Huizenga et al. (2009) explain this by the fact that these generations grow up with multiple information sources through the internet. As a potential remedy to this, digital game-based learning becomes more important. Serious games, where learning is the main objective (Stone, 2005; see Sanchez & Olivarez, 2011, Hong et al., 2009), have the possibility to motivate students at cognitive, emotional and social level.

According to Wouters et al. (2009) these games are more and more used in learning and training complex competencies. However, these serious games are not yet fully integrated; they are not used enough in learning processes. Studying is much more related to 'calm, concentration and research' which is apparently in contrast with the 'fast, pleasure and relaxing' of games. Many educational organizations therefore are rather hesitating to use them in class. The term 'serious game', means though that one learns by playing the games and that learning with games can be fun (Hummel et al., 2011).

Next to the possibility to learn and train complex competencies in a pleasant way, serious games have additional advantages. They have the potential to stimulate cooperation and to facilitate (Csete et al., 2004), but also give the players a better insight in their own cooperation skills and to make them feel responsible for the job and accompanying goals. This leads us to the hypothesis that this kind of games could also give an added value in striving towards better cooperation within a group or between groups. An additional advantage of these games is as well that there is a better observation of details, so that one is more cautious in taking decisions. Because it is considered as being more important to take the right steps, which they can support for 100% (Sanchez & Olivarez, 2011).

Stimulating and promoting cooperation can, according to this research, be seen as a side-effect of these games. Since The Fresh Connection is a serious game, one expects this game to have a positive influence on the cooperation level of each team.

In the literature one can find different taxonomies to classify games. Within the serious game category, TFC can be classified in the subtype "business simulation game", according to the definition Guillén-Nieto and Aleson-Carbonell (2012) provide: it is a characteristic of business simulation games that players have to take decisions and solve problems in simulated business situations.

According to the seven categories that Gros (2007) made, based on different taxonomies in the literature (see Hong et al., 2009) we can place TFC in the group of 'simulations' where "the game takes place in a simplified recreation of a place or situation and a certain goal has to be reached". TFC belongs within this group because it is a simulation of a company that makes and sells fresh juices (Schippers et al., 2012). But the game can also be placed in the group 'strategy games' – this double possibility is one of the disadvantages of this taxonomy, according to Hong et al. (2009). Strategic games have the characteristic that they emulate historical or fictitious situations where players have to find an appropriate strategy to reach certain targets. In TFC the chosen strategy will allow the players to reach the company goal.

Finally, Adams et al. (2012) made a distinction between the different computer games. They mention games where discovering is placed centrally or where the storyline is central or a combination of both. 'Discovering' facilitates the learning process because one is actively involved. The storyline causes events to be coupled together in a better way, which makes it easier to apply what one learned in the real world. We can say that TFC is a narrative discovery game, which means that both methods are being used. For example, in the introduction to the game the CEO of the company asks for help (story) and one needs to find out which strategies have to be applied in this situation (Schippers et al., 2012).

The SCOR-Model

The Supply Chain Council defined the Supply Chain Operations Reference (SCOR) model, as seen in Figure 2 (APICS, 2019). It provides a 'process oriented' view on supply chain management. The supply chain can also be extended to suppliers' suppliers and to customers' customers.

Figure 2. SCOR model overview
Source: Supply Chain Council

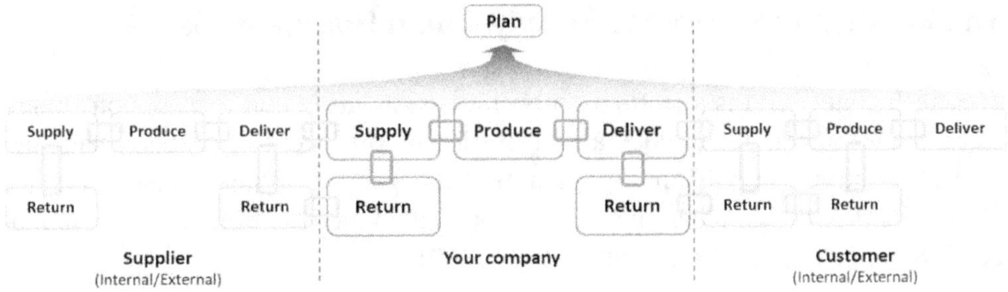

The SCOR-model has been developed to describe the business activities associated with all phases of satisfying a customer's demand (APICS, 2019). The SCOR reference model consists of four major sections (APICS, 2019): • Performance: Standard metrics to describe process performance and define strategic goals • Processes: Standard descriptions of management processes and process relationships • Practices: Management practices that produce significant better process performance • People: Standard definitions for skills required to perform supply chain processes. The model organized around the six primary supply chain management processes of Plan, Source, Make, Deliver, Return and Enable. APICS has also recently launched an app for iOS and Android covering an overview of the entire model (http://www.apics.org/apics-for-individuals/apps/apics-scor-app).

In The Fresh Connection (TFC) decisions regarding these six processes are made and executed by the four managers or VP's.

The Vice-President (VP) Purchasing is responsible for purchasing the components from suppliers. His task is to keep the total cost of ownership under control.

The VP Operations is in charge of the production facilities and the warehouses. His task is to keep the production system flexible and cost effective.

The VP Sales has to negotiate service levels with the customers. He should avoid that the company cannot keep its promises.

The VP Supply Chain Management is the spider in the web, dealing with inventory management to ensure that promises to customers are kept.

By describing supply chains using these process building blocks, the SCOR model can be used to describe supply chains that are very simple or very complex using a common set of definitions. As a result, disparate industries can be linked to describe the depth and breadth of virtually any supply chain. The model has been able to successfully describe and provide a basis for supply chain improvement for global projects as well as site-specific projects. So as a supply chain professional, you can continue your lifelong learning process, guided by the SCOR model.

The Fresh Connection: A Supply Chain Business Game

The fresh connection is a web-based business simulation that focusses on supply chain management. It provides a real life learning process to professionals and students about the supply chain. The aim of this game is to understand strategies within the production chain and is therefore called an S&OP game (Involvation, 2013; Schippers et al., 2012; Cotter et al., 2009).

Figure 3. Screenshot of the fresh connection

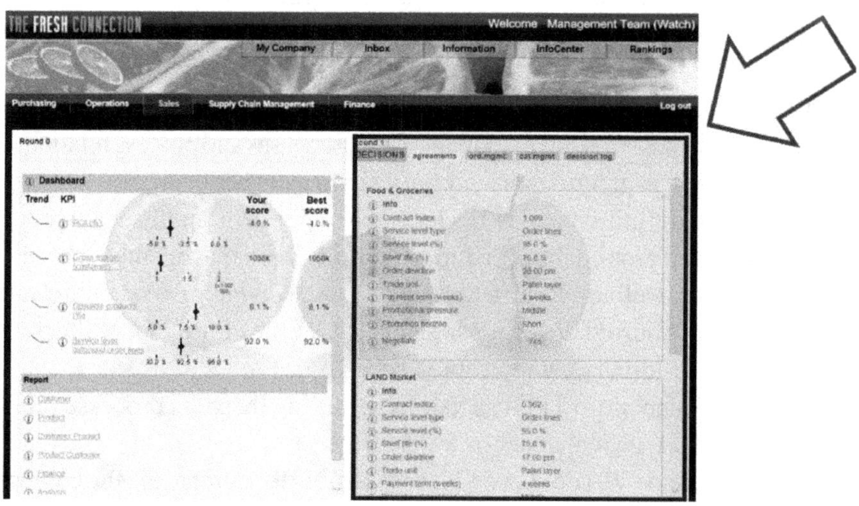

The main goals are to enable the participants to collaborate between different departments and act as a team, and to expand their knowledge on supply chain management. TFC is a fictitious company that deals with production and distribution of fresh juices and is currently facing severe losses. They are not gaining any profit anymore; their inventory levels are too high, the prices for raw materials are going through the ceiling and customers are starting to complain. The goal of the game is to get TFC back on track, by following the best possible supply chain strategy that results in a return on investment (ROI) as high as possible.

The game can be played by teams of professionals or students. Each team consists of four participants, each participant responsible for a corresponding department: purchasing, operations, sales and supply chain. The team has to consider issues like supplier management, demand management, inventory management and capacity management. During the game different teams – each representing a TFC supply chain –compete with each other by applying their own visions, strategies and tactics with regard to purchasing, sales, production, inventory and distribution to the customers.

The game lasts for three to six rounds, each round representing half a year. The competition lasts for two to three days/weeks. The team that scores the highest return on investment (ROI) wins.

In line with the learning cycle of Kolb, the following sequence of activities takes place at each round:

- Analysis of the current situation and problems
- Decision making and implementation of decisions
- Closing of the round and calculation of results
- Reflection on the results, trying to 'conceptualize'

For students and for professionals it is important to structure one's reflections and to conceptualize. This is where the real learning takes place.

At the beginning of the game, there is an information session provided for the participants, starting with a video message from the CEO giving general information about TFC. The CEO states the difficulties that the company is facing: severe losses due to excessive stocks, dissatisfied customers and dissatisfied suppliers. One explains the different roles (sales, operations, purchase, supply chain), along with what is expected of each role, the intention of the game, and the deadlines.

In the course of the game, extra parameters like new products, new challenges and opportunities are added. These new parameters create new supply chain dilemmas. Additionally, extra opportunities are created and new items are added to the game. For instance, in the third round a PET bottle machine can be introduced, and/or a new mixer, with corresponding maintenance contract can be added to the game, or in the fourth round you have to take into account the CO2 emissions of the trucks, etc.

As cooperation is seen as important and a key ingredient to gain competitive advantage, the need for multiple player and role games is large (Hummel et al., 2011). They could have the advantage to stimulate and promote this cooperation (Csete et al., 2004). The players of The Fresh Connection Game, with its motto "Get

Figure 4. Topics covered in each of the three dimensions of supply chain management
Source: Weenk, 2019

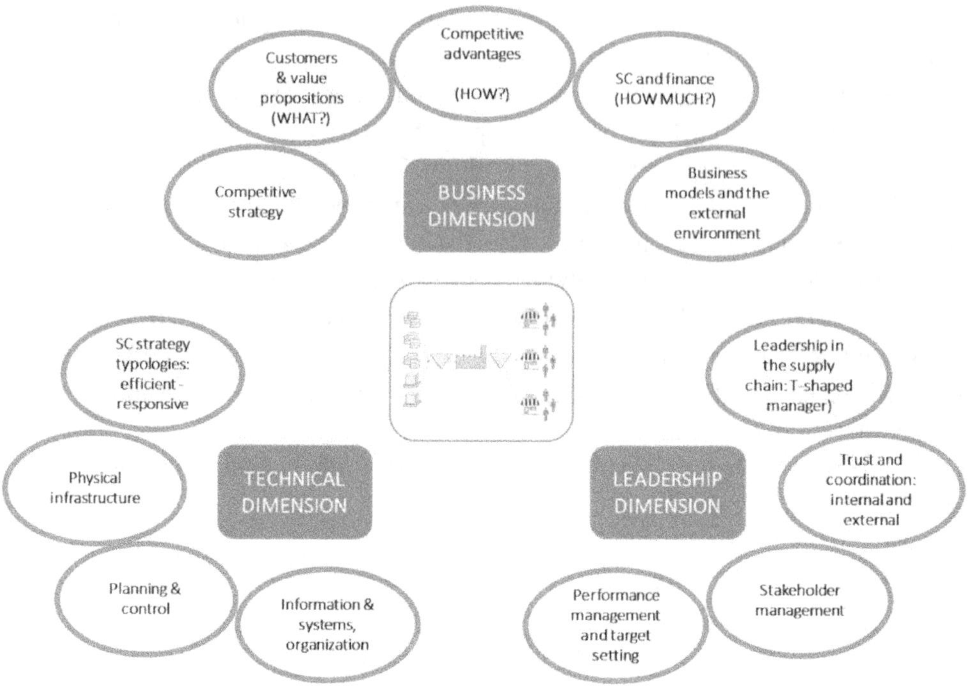

your team connected" (Involvation, 2008), immediately focus on the need to work together as a team in order to win the competition.

Fullerton et al. (2008), would classify TFC in the category 'conflict script' (see Hummel et al., 2011). Conflicts are possible because the four role-players act from different perspectives (from the interests of their own department) to come together to reach a common goal (as much profit as possible). The common goal is represented by the ROI or return on invested capital.

At the end of each of the six rounds (representing three years in the game, each round being half a year) this ROI is calculated and the impact of the individual decisions on the profitability becomes clear. In this way the players get insight in the degree of cooperation needed to increase profitability. This happens in a realistic internet-based environment, where crises, market growth and other incidents happen (Involvation, 2013; Schippers et al., 2012; Cotter et al., 2009).

Different Dimensions in Supply Chain Management

In his book "Mastering the supply chain", Ed Weenk (2019) addresses the different building blocks of the supply chain. To begin with, there is the business dimension, then the technical dimension and eventually the leadership dimension. Figure 4 gives a high-level overview of the topics that can be covered in each of the three dimensions.

Learning About (Supply Chain) Finance

In the game one uses an overall and central key performance indicator, which is also widely used in companies, called the Return on Investment (ROI).

Figure 5 contains elements from both the profit & loss statement as well as from the balance sheet in a company's Annual Report. In addition, the image indicates how supply chain management can have an impact on the different elements (Lambert & Burduroglu, 2000).

Figure 5. Supply chain decisions and ROI
Source: Rushton et al (2017) and Christopher (2016)

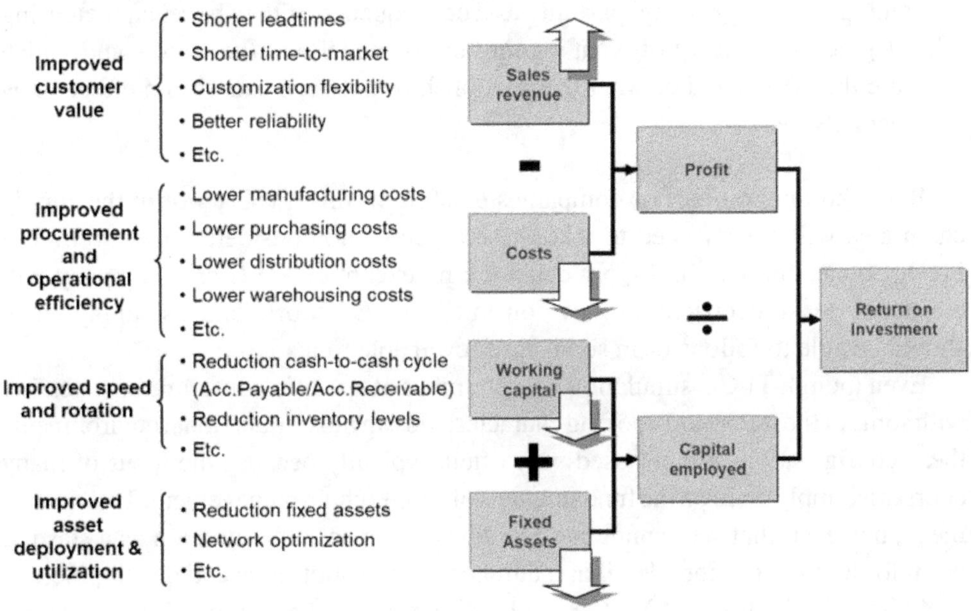

Learning About the Triple Bottom Line of a Company

The triple bottom line is a framework with three parts: social, environmental (or ecological) and financial. Some organizations have adopted this framework to evaluate their performance in a broader perspective to create greater business value. In a specific version of The Fresh Connection dealing with the carbon footprint and even more in The Blue Connection (a set-up around the circular economy in the bicycle sector), one learns about sustainability. Business writer John Elkington claims the phrase "People, Planet, Profit " in 1994. Nowadays profit has become prosperity and even a fifth P, partnership is added. This is well illustrated in the external collaboration possibilities with suppliers and customers in the game.

Learning to Deal with Change

The term VUCA is widely used in business and education. VUCA stands for:

V = volatile. An increasing rate of change at all levels;
U = uncertain. An increasing rate of unpredictability about what might occur;
C = complex. An increasing rate of complexity due to more forces at stake at the same time;
A = ambiguous. Expressing, partially as a consequence of the above, an increasing degree of unclarity of what's going on, of which are the causes and which are the effects and of what the relationships between causes and effects may actually be.

Being an integral part of companies existing in this VUCA world, the supply chain area will clearly need to take these aspects into consideration as well. The Fresh Connection and its supply chain are no exception: TFC as a company will constantly make decisions to move on in the VUCA world and its supply chain should be able to follow, or in some cases even able to lead.

Even though TFC's simulation has a number of specific configurations dealing with some of these forward-looking characteristics, in most educational environments these configurations are not used very often, typically because the focus of many courses is simply more on the fundamentals of supply chain management. This doesn't mean, however, that we cannot use The Fresh Connection's situation as a starting point for looking in more detail at a number of relevant trends and developments.

Following the logic of VUCA, and looking at the world around us in general and the supply chains in particular, change will be the norm. We need to get used to continuous change.

DISCUSSION

The Impact of TFC

Learning as an individual or as a team is very important, because it can lead to competitive advantage compared with other teams/organizations and thus lead to higher profit.

But is this something where serious games and, more specifically, TFC give support? Does the team get more connected whilst playing the game? Does the game have a positive influence on the level of cooperation and does this show in higher profits?

A lot of recent research has been done on the impact of computer games on learning (remembering the presented knowledge or not) and motivation (being more motivated to learn or not). Amongst others Huang et al. (2013), Rondon et al. (2013), Hwang et al. (2012), Wu et al. (2012), Huang (2011), Wouters et al. (2011), Hoffman (2009), Huizenga et al. (2009). There is discussion whether serious games are effective or not and what the reasons are for this (Guillen-Nieto & Aleson-Carbonel, 2012).

Only a few researchers consider the effect of serious games on cooperation in and between groups. Our research looks at the cooperation effect, using The Fresh Connection Game, a supply chain game that tries to improve the cooperation and decisions made within the supply chain (Schippers et al., 2012).

TFC is according to Cotter et al. (2009) a cost-effective way to optimize team performance and in this way to improve company results. But can this game improve cooperation within organizations, and thus reduce the amount and importance of conflicts? If this game would stimulate this, this could accelerate the process towards better cooperation. Allred et al. (2011) have proven in their research that the difficulty of cooperation has to be situated rather inside the organization than between different organizations in the supply chain. And this is exactly where this game focuses on.

There is no doubt that TFC gives a clear view on what the supply chain looks like. If you want to play TFC well, you need to work together as one team. Each individual taking decisions for his or her department just does not work.

The reasons for playing TFC can very diverse, but the main reason is to learn. Learn to cooperate more as one team, learn more about supply chain management, or as a part of a learning process.

Main comments on the game were the following:

1. The game is not 'transparent' enough. It is not clear which decisions led to the obtained result. This is also true in real life, but to improve the learning process

we give feedback on the consequences of the changes of some parameters. As Greco et al. (2013) mentioned "most business games are black boxes", meaning that most of the business games are not transparent at all.

2. The game definitely strengthens the team spirit. TFC shows how different departments have to work together as one team in order to obtain a better result for the whole company. There has to be one aligned strategy that every member of the team has to know and follow. Making decisions as an individual has a negative effect on the game.
3. A simulation of the consequences and a debriefing on the final results and of the departments separately (on the ROI) of curtain actions is an added value to the game. This increases the learning process.

A debriefing session at the end of a Business Game is a crucial element of the learning process of the game. This debriefing sessions helps the participants to convert what they have learned, theoretically and practically, to situations that exist in real life. (Greco et al., 2013).

The Learning Effects of the Game

The game is all about experiential learning and the learning cycle. The learning process starts with the experience, reflecting on what happened, conceptualizing the events and incorporating the findings into the next cycle of experience, either in the game or in your company context.

As a player or in real life, you should assess yourself, try to define concrete goals as well as deadlines. You then define your own (supply chain) learning path and you have to find out what works best for you: self-study with books, linking into web resources from expert associations, subscribing to magazines and newsletters, following groups of professionals on social media such as LinkedIn, finding mentors, reading general business and specific industry newspapers, doing project internships or participate in some courses at university to continuously update your knowledge.

Managers need to be versatile, multi-skilled people. As Weenk (2019) mentions they should be like decathlon athletes, need to perform well on a lot of different disciplines, not necessarily the best at each, but good enough to have a good shot at becoming the overall number one in the tournament". Or as Albert Einstein phrased it "I never teach my pupils. I only attempt to provide the conditions in which they can learn". The Fresh Connection provides you with the conditions in which you can learn, and contributes to your developing a manager's rich skills set and getting ready to manage supply chains or even companies in the VUCA world.

CONCLUSION

Athletes and musicians don't perform without practicing. People in the business world should practice as well. And not by making mistakes, which is a too expensive way to learn.

Simulations such as TFC and recent extensions of the game, The Cool Connection (whose main topic is about supply chain finance) and The Blue Connection (which deals with circular economy), create a virtual practice world that allows teams to test assumptions and experiment. Simulations create the same whole-field perspective as what football coaches and players do when they look at a movie of their game. Each player can step back and look at the whole field. With serious business games such as TFC, one can do the same thing.

This helps accelerating the transformational learning of individuals and teams, based on Kolb's view of learning as a 4-step cyclical model.

Argyris expanded on this model with two concepts: (1) single loop learning and (2) double loop learning. The difference is that in double loop learning one examines one's original mental map and errors in reasoning between the simulation and the person's mental map causes examination of old maps and leads to new ways of thinking and behaving.

For engineers, confronted with growing complexity and an exponentially growing information flow, we are convinced serious business games are a necessary tool to develop the 21st century skills, related to complexity and human interaction in teams. We wish you a lot of learning and fun at the same time.

REFERENCES

Adams, D. M., Mayer, R. E., MacNamara, A., Koenig, A., & Wainess, R. (2012). Narrative games for learning: Testing the discovery and narrative hypotheses. *Journal of Educational Psychology, 104*(1), 235–249. doi:10.1037/a0025595

Allred, C. R., Fawcett, S. E., Wallin, C., & Magnan, G. M. (2011). A Dynamic Cooperation Capability as a Source of Competitive Advantage. *Decision Sciences, 42*(1), 129–161. doi:10.1111/j.1540-5915.2010.00304.x

APICS. (2019). *Supply Chain Operations Reference (SCOR) model.* https://www.apics.org/apics-for-business/frameworks/scor

Cotter, J., Forster, G., & Sweeney, E. (2009). Supply Chain Learning : The Role of Games. *Journal of the National Institute for Transport and Logistics, 10*(3), 32–36.

De Dreu, C. K. W., & Beersma, B. (2010). Fast track reports: Team confidence, motivated information processing, and dynamic group decision making. *European Journal of Social Psychology, 40*(7), 1110–1119. doi:10.1002/ejsp.763

Feng, Y., D' Amours, S., & Beauregard, R. (2010). Simulation and performance evaluation of partially and fully integrated sales and operations planning. *International Journal of Production Research, 48*(19), 5859–5883. doi:10.1080/00207540903232789

Feng, Y., D'Amours, S., & Beauregard, R. (2010). Simulation and performance evaluation of partially and fully integrated sales and operations planning. *International Journal of Production Research, 48*(19), 5859–5883. doi:10.1080/00207540903232789

Guillén-Nieto, V., & Aleson-Carbonell, M. (2012). Serious games and learning effectiveness: The case of It's a Deal! *Computers & Education, 58*(1), 435–448. doi:10.1016/j.compedu.2011.07.015

Hill, G. W. (1982). Group Versus Individual Performance: Are N + 1 Heads Better Than One? *Psychological Bulletin, 91*(3), 517–539. doi:10.1037/0033-2909.91.3.517

Hoffmann, L. (2009). Learning through games. *Communications of the ACM, 52*(8), 21–22. doi:10.1145/1536616.1536624

Hong, J.-C., Cheng, C.-L., Hwang, M.-Y., Lee, C.-K., & Chang, H.-Y. (2009). Assessing the educational values of digital games. *Journal of Computer Assisted Learning, 25*(5), 423–437. doi:.00319.x doi:10.1111/j.1365-2729.2009

Huang, W. D., Johnson, T. E., & Han, S.-H. C. (2013). Impact of online instructional game features on college students' perceived motivational support and cognitive investment: A structural equation modeling study. *The Internet and Higher Education, 17*, 58–68. doi:10.1016/j.iheduc.2012.11.004

Huizenga, J., Admiraal, W., Akkerman, S., & Dam, G. (2009). Mobile game-based learning in secondary education: engagement, motivation and learning in a mobile city game. *Journal of Computer Assisted Learning, 25*(4), 332–344. Doi:.00316.x doi:10.1111/j.1365-2729.2009

Hummel, H. G. K., van Houcke, J., Nadolski, R. J., van der Hiele, T., Kurvers, H., & Löhr, A. (2011). Scripted cooperation in serious gaming for complex learning: Effects of multiple perspectives when acquiring water management skills. *British Journal of Educational Technology, 42*(6), 1029–1041. doi:.01122.x doi:10.1111/j.1467-8535.2010

Hwang, G.-J., Sung, H.-Y., Hung, C.-M., Huang, I., & Tsai, C.-C. (2012). Development of a personalized educational computer game based on students' learning styles. *Educational Technology Research and Development*, *60*(4), 623–638. doi:10.100711423-012-9241-x

Kolb, D. A. (2014). *Experiential learning: Experience as the source of learning and development*. FT Press.

Ku, H.-Y., Tseng, H. W., & Akarasriworn, C. (2013). Cooperation factors, teamwork satisfaction, and student attitudes toward online collaborative learning. *Computers in Human Behavior*, *29*(3), 922–929. doi:10.1016/j.chb.2012.12.019

Lambert, D., & Burduroglu, R. (2000). Measuring and Selling the Value of Logistics. *International Journal of Logistics Management*, *11*(1), 1–18. doi:10.1108/09574090010806038

Lugmayr, A., Sutinen, E., Suhonen, J., Sedano, C. I., Hlavacs, H., & Montero, C. S. (2017). Serious storytelling – a first definition and review. *Multimedia Tools and Applications*, *76*(14), 15707–15733. doi:10.100711042-016-3865-5

Oliva, R., & Watson, N. (2010). Cross-functional alignment in supply chain planning: A case study of sales and operations planning. *Journal of Operations Management*, *29*(5), 434–448. doi:10.1016/j.jom.2010.11.012

Reber, A. S. (1989). Implicit learning and tacit knowledge. *Journal of Experimental Psychology. General*, *118*(3), 219–235. doi:10.1037/0096-3445.118.3.219 PMID:2527948

Rondon, S., Sassi, F. C., & Furquim de Andrade, C. R. (2013). Computer game-based and traditional learning method: A comparison regarding students' knowledge retention. *BMC Medical Education*, *13*(1), 30. doi:10.1186/1472-6920-13-30 PMID:23442203

Sánchez, J., & Olivares, R. (2011). Problem solving and cooperation using mobile serious games. *Computers & Education*, *57*(3), 1943–1952. doi:10.1016/j.compedu.2011.04.012

Schippers, M., Rook, L., & van de Velde, S. (2012). Team supply chain management decisions: Curvilinear effects of reflexitivity and regulatory focus. Rotterdam: Academic Press.

Schippers, M. C., Homan, A. C., & Knippenberg, D. V. A. N. (2012, February). To reflect or not to reflect: Prior team performance as a boundary condition of the effects of reflexivity on learning and final team performance †. *Journal of Organizational Behavior*, *34*(1), 6–23. doi:10.1002/job.1784

Serdarasan, S. (2013). A review of supply chain complexity. drivers. *Computers & Industrial Engineering*, *66*(3), 533–540. doi:10.1016/j.cie.2012.12.008

Suda, L. V. (2016). *NASA's Project Management Leadership LAB*. Academic Press.

Thomé, A. M. T., Scavarda, L. F., Fernandez, N. S., & Scavarda, A. J. (2012). Sales and operations planning and the firm performance. *International Journal of Productivity and Performance Management*, *61*(4), 359–381. doi:10.1108/17410401211212643

Watson, W. E., BarNir, A., & Pavur, R. (2005). Cultural diversity and learning teams: The impact on desired academic team processes. *International Journal of Intercultural Relations*, *29*(4), 449–467. doi:10.1016/j.ijintrel.2005.06.001

Watson, W. E., Johnos, L., Kumar, K., & Critelli, J. (1998). Process gain and process loss: Comparing interpersonal processes and performance of culturally diverse and non-diverse teams across time. *International Journal of Intercultural Relations*, *22*(4), 409–430. doi:10.1016/S0147-1767(98)00016-9

Weenk, E. (2019). *Mastering the Supply Chain: Principles, Practice and Real-Life Applications*. Kogan Page Publishers.

WEF. (2018). The Future of Jobs Report 2018. *World Economic Forum*. Retrieved January 6, 2020 from http://www3.weforum.org/docs/WEF_Future_of_Jobs_2018.pdf

West, M. A. (2012). Creating Effective Teams. In Effective Teamwork: Practical Lessons from Organizational Research (3rd ed., pp. 1–12). John Wiley & Sons, Ltd. and the British Psychological Society.

Wouters, P., Spek, E. D., & Oostendorp, H. (2011). Measuring learning in serious games: A case study with structural assessment. *Educational Technology Research and Development*, *59*(6), 741–763. doi:10.100711423-010-9183-0

Wu, W.-H., Hsiao, H.-C., Wu, P.-L., Lin, C.-H., & Huang, S.-H. (2012). Investigating the learning-theory foundations of game-based learning: A meta-analysis. *Journal of Computer Assisted Learning*, *28*(3), 265–279. doi:10.1111/j.1365-2729.2011.00437.x

Chapter 10
Using Gamification and Serious Games to Design a New Curriculum

Kutay Tinç
Istanbul Technical University, Turkey

Meltem Gülçin Karadayı
Istanbul Technical University, Turkey

ABSTRACT

Using game elements in class to support the participation of students in learning or designing games that can help educators teach certain subjects more efficiently has been a popular topic in recent years. The former is a matter of gamification, which refers to the application of game elements to other activities so that the activity becomes more engaging or interesting. On the other hand, the latter is about designing a serious game, which can be defined as a game with an explicit and carefully thought out educational purpose. In this study, focused on merging the use of gamification and serious games for a specific engineering course, the authors discuss how the curriculum for this course should be designed so that both sides of the spectrum are facilitated. An application of this union is given with a survey showing the reaction of students to the gamified curricula integrated with a serious game.

DOI: 10.4018/978-1-7998-2562-3.ch010

INTRODUCTION

No matter how a learning or a teaching process is designed, there are always subjects that some students are having problems with and are discouraged to study on. This can either be because of the students' lack of interest in the subject or the subjects' difficulty itself being too high for those students at that time. This, coupled with the students of the digital age who have been raised in constant contact with a digital world, makes the job of teaching certain hard subjects to them an almost impossible task. Even simple subjects that require constant attention can turn into hard subjects when the attention of the students wander, and with technology so integrated in our lives this is only natural. Seeing students play mobile games in classes or even a few of them snoring on the background has become a natural phenomenon. This trend has resulted in lower performances in all grades and continues to do so even in higher education. To battle this trend, many different approaches are being tried and one these approaches is using game elements in lessons.

As the digital age slowly creeped in with the start of 21st century, video games have started to become more and more accessible and hence more popular among all age groups and genders (Entertainment Software Association, 2015). With the evolution of streamers and streaming platforms like Twitch and the rise of Esports, video games can easily be considered the most promising entertainment media of the future, whether we play them or watch them.

This influx of popularity coupled with the fact that video games have a high motivational potential (Garris, Ahlers, & Driskell, 2002; Gee, 2007; Hense &Mandl, 2014; Przybylski, Rigby, & Ryan, 2010; Rigby & Ryan, 2011; Ryan, Rigby, & Przybylski, 2006; Yee, 2006) naturally gave birth to the idea of gamification.

Although the first utilization of the "gamification" term is claimed to have been made in 2002 by a game designer called Nick Pelling (Werbach & Hunter, 2012), Deterding, Dixon, Khaled and Nacke (2011), claims "Gamification" term was not known at all before the various conferences and popularization of it in 2010 and was characterized by the digital media industry in 2008.

Serious games, on the other hand, refer to games with educational purposes. These games do not have to be digital games, but for our purpose we will consider digital serious games throughout this chapter, and digital serious games fall under the purview of GBL. GBL, is the abbreviation for Game-based learning, which means the usage of games in academic world. (Wiggins, 2016).

This chapter is structured as follows: After the introduction part gamification will be discussed thoroughly with all its aspects and use cases, especially in education, then serious games will be analyzed and after the part about how a general curriculum can be designed, the authors will be focusing on merging the use of Gamification and Serious Games for one course for engineering education and how a curriculum for that course can be designed that will be able to make use of both sides of spectrum. An application of this union which is being tested in a university course is also given with a survey showing how the students reacted to the gamified curricula with an integrated serious game.

BACKGROUND

Many researchers have defined gamification as the implementation or usage of game design elements or game mechanics in non-game environments to foster motivation (Deterding et al, 2011), while this notion is true, different aspects of game mechanics can be used in a diverse number of applications which can change how gamification can be applied to a certain environment (Sailer et al, 2017). One of the main purposes of applying game elements into non-game areas is to attract users and keep them engaged for a longer period (Kasurinen & Knutas, 2018). Studies have also stated that learning occurs naturally while playing games (Deif, 2012), hence gamification, which turns non-game environments into game like experience, can be considered as a tool for learning.

Research on gamification has shown that it has already been implemented in various context such as: business (Arai, Sakamoto & Washizaki, 2014; Fernandes et al., 2012), crowdsourcing (Liu, Alexandrova & Nakajima, 2011; Mekler, Brühlmann, Tuch, & Opwis, 2015), data-collection (Downes-Le Guin, Baker, Mechling, & Ruyle, 2012), health (Jones, Madden, & Wengreen, 2014), marketing (Hamari, 2013, 2015), social networks (Farzan & Brusilovsky, 2011), environmental protection (Gustafsson, Katzeff, & Bang, 2009), and education (Landers & Landers, 2014; Shi, Cristea, Hadzidedic, & Dervishalidovic, 2014).

When these implementations are considered, the purpose of gamification can be summarized in three parts as: striving to engage people to a goal achieving process, motivating people to develop skills, or fostering attitude shift, by applying game design principles to non-game environments (Deterding et al, 2011, Kapp, 2012).

The premise of gamification is mostly based on game elements and motivation. When it comes to game design, the aspect of motivation is of utmost importance and that makes gamification a perfect learning method. As the topic or actor of a game is always a person, game designers always need to be interested in psychology and behavioral sciences. It is important to know how the player can react or respond to game elements, mechanics or even the story. Any design element should be thought out thoroughly and modelled for the specific purpose the game or gamified process wants to achieve.

Studies show that people become more productive during games (Kim, 2012). One of the most crucial concerns teachers have for their classes are low participation in class and the absence of interest of the students (Lee & Hammer, 2011). This problem, taking place in many classes, can be solved by using the effect of the games on people.

Games mostly have some objectives for players and objectives tend to focus people and get them motivated. Changing or reprogramming attitudes as a problem-solving method can be intended with psychological practices and elements of a game (Everson, 2015). 75% of gamification consists of psychology (Zichermann & Cunningham, 2011). The rest of it are the technological parts, which includes the coding of game elements and any interaction technologies that will be used. Zichermann & Cunningham also explains that, motivation can be divided into two groups: intrinsic and extrinsic motivations.

Intrinsic motivation can be defined as a continuous and internal drive such as voluntariness, mastery and willingness. On the other hand, extrinsic motivations derive from any other external motivator such as money, badges, trophies, advancements etc. They are an effective way to motivate people to accomplish basic duties but not adequate for any task that requires some sort of innovative approach (Zichermann & Cunningham, 2011; Kumar, 2013). Gamification practices mostly focus on external motivators like badges, trophies and achievements.

Gamification in Education

Contrary to the general belief, games are not developed just for children. Markopoulos, Fragkou, Kasidiaris and Davim (2015) have stated that games are popular among adults as well as the young generation and gaming is not gender biased. These propositions can be derived from the fact that average age of the gamers' is 37 in USA and 48% of the American gamers are females. The gender percentage of gamers are also equaling out fast in other countries such as France, Italy and England (Markopoulos, Fragkou, Kasidiaris and Davim, 2015). Hence every age group and gender can enjoy games as much as any other. Games are a very common part of almost every culture. And as we know that games are for everyone regardless of age and gender, that makes games an appropriate option for general educational use.

Gamification, just like education, is multidisciplinary. They are compatible and can complement each other. Teaching by gamifying is utilized as a popular disciplining way not only in families but also in schools or universities during recent times. The most obvious sign of it, surely, is the grade system that we use, which can be classified as a reward system.

Following that train of thought, a study was conducted on an operation research course for business students (Dias, 2017), where two distinct classes were created.

The first class used a gamified curriculum and the second one was a normal class without the use of gamification. The area covered by this study was slightly more extensive as other than observing effects of gamification on students, it also observed the effects of gamification on the lecturers. The outcomes of this research had positive impacts on both the students and the lecturers. For the students whose lessons were based on gamification, their success and interest to lesson were higher compared to the non-gamification curriculum-based class. For the lecturers whose lessons were gamified, their commitments to the class were higher. So gamified curricula positively affect not only the students of the course, but also the lecturers as well. This opens another venue for research, where the effects of gamification can be tested on lecturers to find an optimal model that maximizes both students' and lecturers' welfare in the class.

Educational gamification cannot only be limited to schools and universities. Markopoulos et al. (2015) have also mentioned that 2.6 million games have been downloaded in Germany in 2011. Furthermore, almost half of the employees and more than 60% of CEOs and CFOs play games at work in Germany, also according to a survey conducted by Ipsos for Saatchi & Saatchi, 55% of the employees would be willing to work in an organization that uses games to boost performance (Ipsos, 2011), both of which show the penetration power of games and thus it would be logical to apply the gamification mentality to business education as well as traditional education.

As Lee and Hammer (2011) have pointed out, gamification in education is used for shaping and changing behaviors of students. This is possible with its three major intervention area. These can be listed as follows: "cognitive", "emotional" and "social".

Cognitive: Complex rule systems, which can be solved mostly by trial and error, lead gamers to actively explore the game mechanics and try to figure out the next move. These duties which should be matched with his/her level help them to stay engaged (Lee & Hammer, 2011). Multiple paths should be created to allow players to create their own moves and paths until the end with objectives that get harder gradually along the path (Locke & Latham, 1990).

Emotional: Games are a medium where people can experience strong feelings; such as enthusiasm, anxiety, curiosity, joy, scare, anger and disappointment. Short feedback cycles of games encourage people see their failures as a part of getting better and achieving success in the end, instead of giving up (Lee & Hammer, 2011).

Social: They also explained that one of the most important characteristics of games is finding out your other successful roles and identities at a safe environment. Gamification helps people find their potential by giving them meaningful and effective roles (Lee & Hammer, 2011).

Some psychologists believe that games are the way of working for young generation. Lee, Luchini, Michael, Norris & Soloway (2004) explained the research about "Learner-Centered Design" (LCD) to support activities by using technology, curriculum and personal computers. LCD is developed for customization of education and engagement of students by recognizing their needs. LCD method also includes some elements such as; avatar, level, score and ranking. Even using LCD, we can understand why it is important to use a method like gamification for motivation. Even if it was not named as gamification, the mentality is the same.

Gamification is becoming a very important part of education, but it is not the only one. Serious games are being coded and served for use in companies, classrooms and other venues to help people learn new things while playing games.

Serious Games

Although the use of games to gain skills has a longstanding history, the serious game term was first used by Clarke C. Abt in 1970 (Wilkinson, 2016). Abt describes serious games as games played for an intent rather than entertainment (1987). However, serious game term is used for generally digital games that have an educational purpose today (Crookall, 2010).

Figure 1. The relations between serious games and similar educational concepts (Breuer & Bente, 2010)

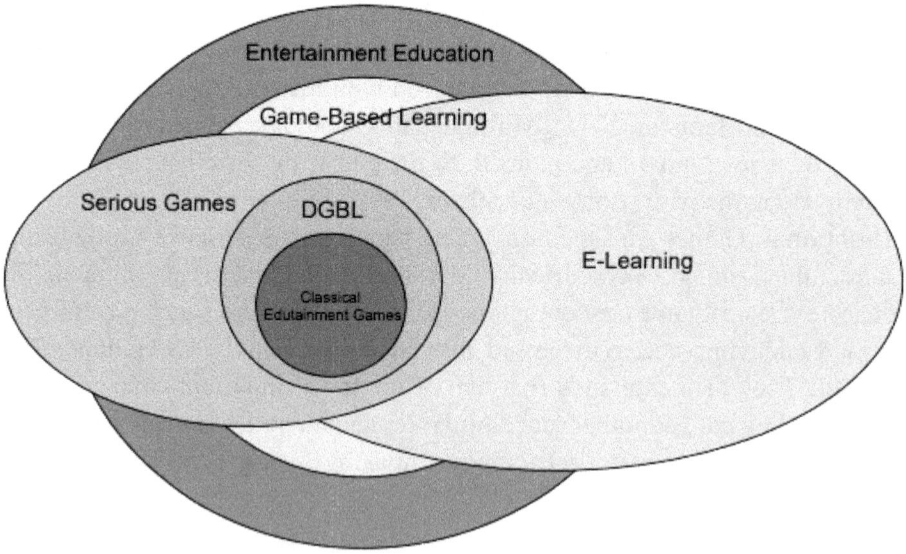

Besides serious game, there are also different concepts that aim to facilitate the learning process. Breuer and Bente formed a scheme to show the position of the serious game between other concepts (Figure 1).

Aside from showing the relations between different educational concepts, they have stated that entertainment education involves all kind of tools that combine education and entertainment, while game-based learning comprises the use of any kind of game for education under this concept. They also mentioned that serious games can be used as an e-learning method. The e-learning method is another new trend for people that want to develop their skills no matter what their level or grade is. This method may also include various gamification design elements to increase its efficiency, however, e-learning does not need to be fun and serious games have different application areas from learning like rehabilitation, marketing, etc. They have also explained that digital game-based learning consists of digital games with the main purpose of training as a part of serious games (Breuer and Bente, 2010).

There are several advantages that serious games provide to enhance the learning experience. Dijk et al (2015) studied essential learning elements and compared serious games and presentations in terms of these elements as seen in Table 1. They also pointed out that both learning methods include determined learning elements, however serious games and presentation have different ways of providing them. While serious games insert learning elements into the game and have a more learner-centred approach, presentations give the information to learners directly and make them more passive in the learning process.

Table 1. Mapping serious games and presentations on learning elements (Dijk et al, 2015)

Learning Elements	Serious Games	Presentation
Reflection	In game feedback on decision made & Debriefing	Presenter – Audience Interaction
Challenge	In game goal setting & Competition	Question the Audience & Provide a case
Control	Players can make their own choices & Game adapts to the player	Address content based on audience preferences
Goal Setting	Present learning goals at the start & In game goals	Present learning goals at the start
Failure	In game experience of problems	Present a problem & Ask questions to the audience
Practice	Practice while playing	Stimulate active thinking by showing examples
Interaction	In-game interaction & Interaction among players	Involve the audience

According to Garris et al (2002) characteristics of a digital game can be grouped in six dimensions; fantasy, rules & goals, sensory stimuli, challenge, mystery, and control. They emphasized that fantasies give users opportunity to act in different situations without having any impact in real life. Through fantasies, concepts in real world can be replaced in games as a phenomenon and can be experienced in different aspects (Malone and Lepper, 1987). These functions of fantasies give users the opportunity to see the results of their actions freely. Driskell and Dwyer (1984) stated that fantasies enable users to be absorbed with full concentration in game. They stated that a serious game that intrinsically combines learning content with its fantasy may have the ability to provide motivation to learn the relevant content. Another dimension of games that stated by Garris et al. (2002) is rules& goals. They mentioned that games have specific rules to reach certain goals and feedback is provided to users based on how far they need to progress to achieve these goals. They estimated that users tend to reduce the distance between goals and their current positions in the game, which leads to motivation and performance. Besides, they highlighted that games present experiences to users that they cannot live in real life. They explained that sensory stimuli like sound effects and dynamic graphics catches the attention of users. Moreover, Rieber (1996) stated that an optimal level of challenge encourages learners to reach goals. He maintained that when faced with a challenge that cannot be solved in an easy or hard way, a learner will be motivated to find the solution. Furthermore, Garris et al. (2002) explained that mystery is another factor that games provide to support the learning experience. They maintained that the mystery created triggers the curiosity of users and keep their attention to see what will happen next in the game. Additionally, they remarked that games also give students the opportunity to choose their own strategies, and manage the course of the game. They asserted that the sense of control increases the motivation and enhances learning.

Qian and Clark (2016) pointed out, while the target groups of studies conducted on game-based learning related to 21th century skills development consist of all age groups, most of these studies, which constituted 30 percent of them, examined the higher education students as the focus group. They also stated that game-based learning increases the effectiveness of learning processes when they are designed by combining learning theories and game elements appropriately. Subhash and Cudney (2018) expressed that gamification and game-based learning in higher education have lots of application areas from science to communication, while specially business and science studies are increasingly applied. They also mentioned that gamification and game-based learning in higher education have lots of advantages for students and teachers like increased engagement, motivation, and performance. Barr (2108) declared that higher education students have a positive view of using commercial video games to gain graduate competencies, such as communication, resourcefulness, and adaptability. Subbash and Cudney (2018) also stated that gamification and game-based learning in higher education typically include badges, such as achievement badges, rewards, and virtual trophies; the points such as experience points, bonus points, and gained points; and the levels such as quests, missions, and challenges as game elements.

Hauge et al (2012) investigated three serious games on how serious games have enabled engineering students to gain the necessary competencies for engineering practice: COSIGA, Beware Game, and Set Based Simultaneous Engineering Game (SBCE). COSIGA is a simulation game played with five players which makes it available for players to experience five disciplines in New Product Development: design, manufacturing, marketing, project management and purchasing. Beware game is a multiplayer online game which can be used to understand risk management based on supply networks in companies. And SBCE is a teamwork game that shows to players that using SBCE concept increases time and cost efficiency on airplane production. The examination based on these games demonstrates that although players' previous knowledge was an important factor, games enabled that players to use theoretical knowledge they have learned and increased the ability of players to cooperate. Braghirolli et al. (2016) suggested that serious games are beneficial as perambulatory activities for first class industrial engineering students. They developed a game that simulates the competition between different juice companies in the same market by revealing the factors needed to demonstrate the complexity of the subject. The results of the game pointed out that serious games increased the motivation and engagement of first-class industrial engineering students. Bellotti et al. (2014) investigated the current serious games about entrepreneurship concept that could be used for higher education in the market. They stated that while lots of company management simulations are available on the market, there are relatively less serious games in regards to innovation and motivation. They also mentioned that fun is a missing factor in current serious games. Dobrescu et al. (2015) stated that video-games could be used as a substitute for traditional teaching methods when considering cost efficiency and quality, based on an experiment with a game designed about microeconomics among undergraduate students. Pasin and Giroux (2011) evaluated the usefulness of HECOpsim simulation game to teach necessary skills for operation management to undergraduate business administration students. The game consists of linked spreadsheets that allow students to decide on the sales forecast, the quantity of raw materials to be purchased, the number of workers to maximize their profits in a manufacturing firm. They expressed that introducing game-based learning to traditional teaching methods enhance complex decision-making skills of undergraduate students by endowing those students with experience. Coller and Scott (2009) reported that game-based learning increased the time spent by students and students' level of understanding compared to traditional teaching methods, based on a course of numerical methods that included a serious game that allowed mechanical engineering students to learn automobile physics. Smith and Chan (2017) asserted that using game-based learning to teach programming increase the motivation of engineering students according to the results of Space Race game, which is a tablet game that incited competition and collaboration, through questions that should be solved with a time constraint.

Curriculum Design

Gamifying a course is by no means an easy task. It requires creativity and knowledge of the course above all else but certain degree of skill in game design is also helpful.

A gamified course would make good use of interactive in class activities where students can be pitted against each other, or made to work cooperatively, without making the activity too hard or complex for any one of them. The most important aspect of the curriculum should be to not let the students feel stressed or bored, which would lead to students losing focus.

This also requires some degree of, for lack of a better word, showmanship. The lecturer should be in control of the class the entire time by interacting with the students and leading them. In class activities should be tailored to match the speed and style of the course. For example, a history lesson can contain some roleplaying moments for the students, whereas a mathematics lesson can feature a game show that requires a mobile app to enter.

All courses can have different designs but the onboarding technique of mobile games can be used to draw students into the lesson and keep them interested. Onboarding helps players learn how to navigate the game and it can be done either passively with a walkthrough, or actively with a hands-on approach. In the hands-on approach the player is usually left floundering in the game till they learn how to navigate by themselves. Although this approach is good for some games, for education purposes it usually leaves students confused and stressed.

A good approach is to start with the easier subjects, where the lecturer helps the students a lot and raise the difficulty slightly as you keep going and let them navigate by themselves. Then, when the students seem to get stuck, lower the difficulty by switching to an easier subject or by giving the students appropriate hints. This of course is easier said than done and will require a lot of testing to achieve.

Such a designed course will be given as an example for the union of gamification and serious games in the next part of this chapter.

The Union

Gamified curriculum usually refers to in class activities that involve awards like achievements or badges given to students for completing certain tasks like being the fastest student to solve a specific problem or even being able to solve a problem. While this method and its similar counterparts are encouraging and viable, we believe that more can be done to upgrade our curricula on most courses.

Both Gamification and Serious Games have been analysed in different ways by academicians, from saving energy scenarios (Orland et al, 2014) to management education for sustainability (Sharma, 2017). Some papers published focus on how a course can be gamified (Müller et al, 2015) or how Serious Game should not be designed (Malliarakis et al, 2015).

What we will be discussing here is already being tested on one of the courses in ITU Industrial Engineering Department and students enrolled in the course have given positive feedback about their learning experience. The details of the class, how the proposed model is applied there and the results in the eyes of the students of the course will be given at the end of this part.

Most university courses have one or two Midterms, and a Final Exam. Some courses have homeworks, group projects, quizzes or similar grading mechanisms as well. To achieve a gamified curriculum, we need to have some sort of in class grading mechanic like in class activities or weekly quizzes.

Designing these activities and quizzes as part of a larger serious game will require them to be digital as opposed to pen and paper, hence they can be graded automatically by an artificial intelligence, if the questions need written answers, or simply by true / false counts. Digitalized in class grades will also allow us to give out meaningful badges and achievements as well. Any in class action can turn into different types of rewards to be handed out to the student. And the variety of these rewards is a boon for any game designer.

These in class activities can be done with a mobile application that can be written specifically for the course which can transfer data directly into the serious game which the course will use. So that any in class activity can be kept track of and achievements or points earned by students can be synchronized with the serious game instantly.

For in class activity achievements part, let's assume that any student can earn the achievement "Speedy Gonzales" for being the first to answer a given question. This achievement does not have to be earned with every question, one or two questions per week might be an ideal limit depending on the class population. Also, this achievement can have a time interval to allow more than one winner. This interval can be designed as follows: When the first student to answer the question correctly emerges a timer for a set number of seconds start, a fair choice might be 10 seconds, and all students who answer correctly before the timer ends get the achievement as well.

Figure 2. Currency conversion into different resources

Currency Type		Resource Type
Achievement - 1	–>	Manpower
Achievement - 2	–>	Oil
Correct Answers	–>	Money

Let's also assume that getting 10 questions correct in a row gives you the 'Winning Streak' achievement. This achievement can be earned by all the students if all of them can answer 10 questions correctly in a row. This achievement can be designed to not include questions already used for the previous 'Winning Streak', so 20 questions can only give this achievement twice at most instead of 11 times. This would make the achievement a bit fairer for students who can't always answer everything correctly.

Let's add the number of correctly answered questions itself as reward points, too. We now have 3 different *currencies* we can use for further development. These currencies can be earned by all students but while it is probabilistically possible, it is not probable that everyone can earn all of them at the same time. While all students can answer a question correctly, being the fastest ones would only belong to a select few.

Leaving these currencies aside for now, let us focus on the serious game part, which can be implemented as a project for the course.

The idea behind a course project is to help channel students into doing research or study the subjects they have seen in the course and implement them. A serious game designed as a course project can have different mechanics ranging from simple text-based interaction to more advanced graphical interfaces or even VR components. All of which can have students apply their course knowledge and have fun doing it.

Depending on the type of the course, the serious game can be designed as a strategy game, a puzzle game, a role-playing game or a simulation. Other types of games are viable too but this is up to the skills and imagination of the game designer. Most games feature a resource, but for those games that lack even the simplest resource, time, the designer can try using hints or help mechanics that can be linked to the in-class activity currencies.

Again, to give an obvious example, let's assume the serious game is modelled like a strategy game. In strategy games players have various resources they can use to achieve specific objectives. These resources can be manpower, types of money, strategical resources, time, buildings, armies and the like. Then any currency earned during in class activities can be used in the serious game after it is converted into strategy game resources. As seen in Figure 2, each type of currency can be converted into a different type of resource, but that of course is not a necessity. Attaching different weights to the currencies and converting them all to a single type of resource

Figure 3. Currency conversion into same resource

Currency Type		Resource Type
Achievement - 1	–>	25$
Achievement - 2	–>	100$
Correct Answers	–>	2$

is also possible. An example of this application can be seen in Figure 3.

This conversion and its applications are the mechanic that makes the union of gamification and serious games in education possible. As any type of serious game can have some sort of objective that presents a challenge to the player, any reward earned during in class activities can be transferred into the serious game seamlessly. This is not a matter of 'can' but a matter of 'how' and implementing it comes down to the skill of the game designer.

Skill of the game designer defines how well the game is fleshed out and how interesting or teaching the game is. As students that take interest in the game will spend more time with it and learn new things trying to crack it, making a fun game that properly teaches the subjects that it was designed to teach will be the burden of the designer.

When both in class activities and the course project are a part of the grading mechanism, and the bonus resources earned from in class activities effect the performance of the student in the serious game that is the course project. This enhances the grading effect of the in-class activity, which is another motivator for the student to participate as well.

Application

In 2017, the new head of the ITU Industrial Engineering department announced that he wants to integrate gamification into the curriculum of all courses and formed a team for this purpose. This application is one of the products of the work done by members of the team.

This union of gamification rewards from in class activity and serious game resources is being tested in the course *END369E Game Theory* of Istanbul Technical

University Industrial Engineering Department. The project is in its second year as of now. The method consists of in class digital economic games and the game project with 12 weeks of turn ins.

Game Theory course, naturally, is the perfect environment to apply this model as the course itself teaches how to strategize in interactive decision-making situations and can easily be modelled as a multiplayer game.

In class games used during the course can be found in https://economics-games.com/ where many interactive games to teach various types of economics subjects to students already exist online [Gruyer and Toublanc]. These games pit students against each other or force them to work cooperatively to achieve some goal. Even when they are cooperating as a team, they are competing with other teams, hence a reliable strategy is of utmost importance.

Students scoring top points in these in class activities gain 'Relocation Tokens', 'Upgrade Tokens' or 'Hype Tokens'. These tokens can be used as an advantage in the serious game project. Their effects will be made clear after the game is explained in detail.

The serious game project consists of weekly turn ins and is a somewhat complicated strategy game with the purpose of maximizing the total money a player has at the end of 12 weeks. The game itself is setup as follows:

Each player can see a 35 *x* 35 map of the entire region with 20 cities or markets where they can sell their homogenous goods and they have to pick a square out of 1225 squares to build their factory. This factory can be placed on a city, or other players, albeit unknowingly, can place their factory on your square as well. This location sets their transportation costs along with Euclidean distances to markets:

$$\text{Euclidean Distance} = \sqrt{\left(Factory\,X - Market\,X\right)^2 + \left(Factory\,Y - Market\,Y\right)^2}$$

So, the location of the factory is the first strategical decision the students have to make even before the game starts. This decision is not only based on the distances but also on market properties. Each market has different *'Ranking'* and *'Growth'* values. These values are coefficients that determine how a market is affected from pricing and marketing strategies. One couple of coefficients affect the ranking order of students while the other couple of coefficients affect how much the market grows, or shrinks, before the next turn starts.

Students have to decide on the price and marketing investment for each market separately and this will affect their ranks on that market. As each market will be affected differently from price and marketing investment, they will require a different strategy.

Each student will be ranked for their pricing and marketing strategies where the cheapest price and the greatest marketing investment each turn get the lowest rank respectively for their price - marketing couples. Then these ranks will be multiplied by ranking coefficients of the market and their total rank value will be obtained. This total rank value is called the final rank and when ordered ascendingly these values will determine the market leader and its followers for that turn, which in turn will be used to calculate who can sell their goods at what price in which market. The market leader sells the amount of product he/she has sent to the market as long as that amount is less than the total demand of the market. Then the rest of the students sell their products according to their rank as long as there is still demand in the market. An example of this procedure can be seen in Figure 4.

Figure 4. Goods sold by students according to their ranks

Rank	Player	Goods Sent	Demand Left	Goods Sold
1	Player 4	300	600	300
2	Player 13	150	300	150
3	Player 11	50	150	50
4	Player 2	200	100	100
5	Player 21	150	0	0

During each turn, students have perfect information about all previous turns, so they know what other students did and can strategize accordingly. Their strategies are made up of actions they can take; producing goods up to their factory limit, sending goods to markets of their choice up to their available quota which is the sum of inventory they had from last turn plus what they have produced this turn, choosing a different price for each market they have sent goods to, choosing the amount of marketing investment they want to make for their goods in the markets they have chosen, increasing their factory production limit for the next turn and using a token they have won from in class activities.

As mentioned before, students have three types of tokens they can earn and use;

Relocation Token: This token instantly relocates their factory to a new square of their choosing. The relocation takes effect instantly and before any other action during the turn, hence all transportation costs will be calculated after the relocation. Students can have only one of these tokens saved up, any new token they win will not be stored.

Hype Token: The Hype token is used to boost a student's popularity for one turn, making his marketing investments more powerful. When a Hype token is used, any marketing coefficient used as a multiplier for calculating the final rank of the relevant student will be halved, effectively lowering the final rank value of the student and giving them a chance to get a better final rank order. Students can have up to two of this type of token but will only be able to use one per turn.

Upgrade Token: The upgrade token can be used in different ways. First one is a free production capacity upgrade where the production value will receive a permanent 10% boost. This boost will happen before the turn starts hence the new limit can be used to produce more goods. The second one is to lower the transportation cost permanently by 25%. This will effectively cut down on costs and make less expensive prices viable. The third and last way of using this token is to lower the production costs permanently by 10%. Each usage is multiplicatively cumulative hence the costs cannot be 0. Three tokens of this kind can be saved for later use.

It is obvious that strategically these tokens will be important, hence winning them from in class activities will be an extrinsic motivational trigger for the students, which in turn is expected to make them show more effort during the course.

The serious game that was played by the students can be analyzed in the website http://bit.ly/TheGame2019Fall. As players make their moves, rounds are updated and the final ranking table is available for examination.

Application Results

The success rate of the application of the union mentioned before was observed through a survey done to the students taking the course in 2018. In this survey students were asked to rate four statements from one to five scale in this order:

1. Strongly Disagree
2. Disagree
3. Neither Agree nor Disagree
4. Agree
5. Strongly Agree

The first statement that students were asked to rate was "*S1 = The fact that the project is in the form of a game makes learning easier*". As seen in Figure more than 85% of students believe that having their term project in the form of a game made learning easier.

The second statement that students were asked to rate was "*S2 = Gamification makes it easier to study outside of school hours*". This statement targeted the study habits of students and the results seen in Figure are quite positive as about 70% of students agreed with the statement.

Figure 5. Scoring of Statement 1 "The fact that the project is in the form of a game makes learning easier".

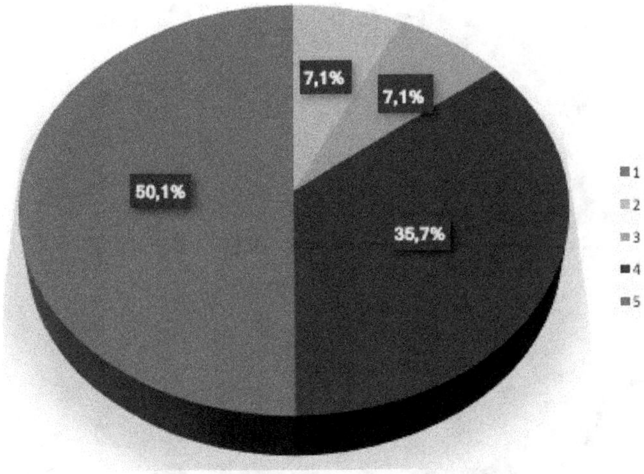

Next statement tested the motivational effect of having a game as a term project with the statement: "*S3 = The fact that the project is in the form of a game is motivating to make efforts to learn*". This statement, again, showed great positive feedback.

Figure 6. Scoring of Statement 2 "Gamification makes it easier to study outside of school hours".

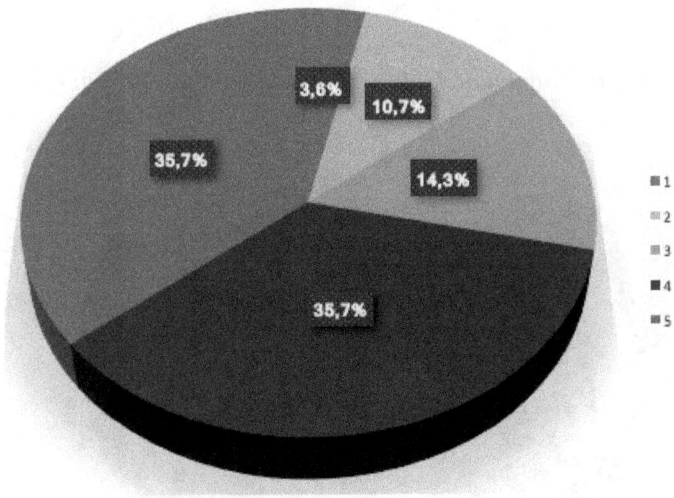

Figure 7. Scoring of Statement 3 "The fact that the project is in the form of a game is motivating to make efforts to learn".

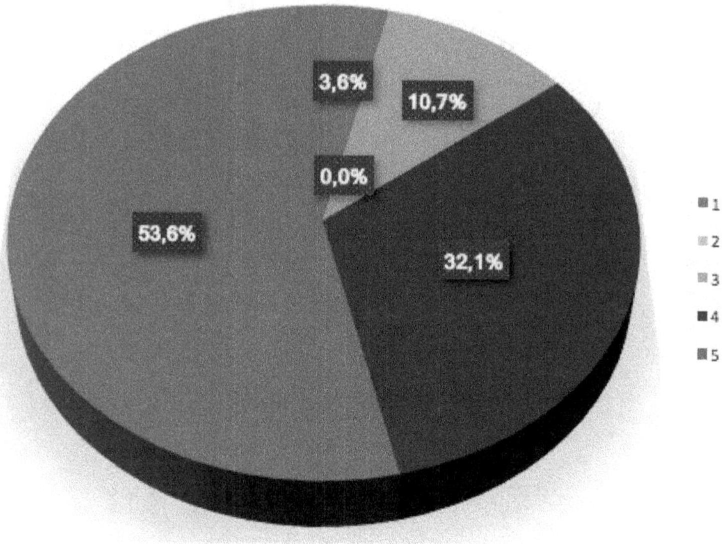

In addition, the last statement the students were asked to rate was "*S4 = I want to use gamification in other courses*".

This statement shows 82% of students want more gamified curriculum or GBL applications in their courses.

Figure 8. Scoring of Statement 4 "I want to use gamification in other courses".

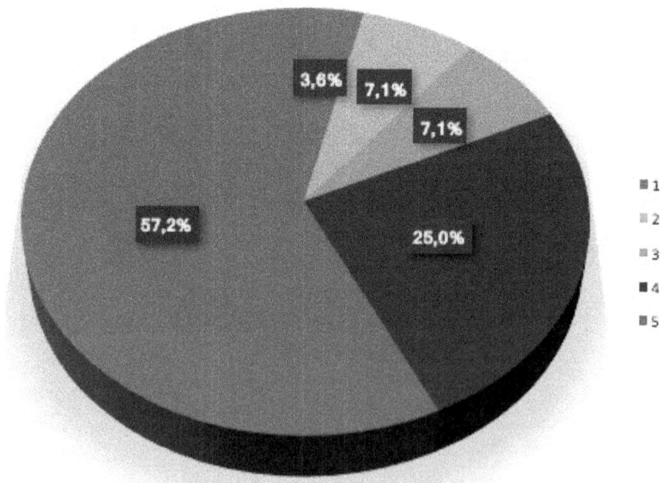

These ratings on the four statements show that most of the students think that gamified curricula and GBL applications will be useful for their learning process and they want to see more similarly structured courses.

Future Studies

Further study can be done by designing an all-encompassing curriculum for consecutive lessons where achievements and points earned in one course can be transferred to other courses up to a point. This mechanic can even be applied to all courses in a university program and a student can be expected to earn certain achievements to become viable for certain elective courses or even for graduation. This would encourage students to study more with a clear goal in mind and further increase their effectiveness.

CONCLUSION

Traditional teaching methods can only be effective up to some point. Compared with the traditional education methods, gamified education methods are more consistent and help students to get better results faster. These propositions have been studied by different academicians and tested in various studies, but just the gamified curriculum itself is not enough when we can devise a way to further motivate the students and even educators by adding a serious game application to the courses.

Gamifying in class activities via a mobile application helps keep track of student achievements as well as motivating them to participate more. The achievements or points earned by students in these in class activities can be transferred to the serious game that is designed as a project for the course and used as a resource in there. This not only links the in-class activities to the project, but also increases the grading effect of said in class activities.

As gamification penetrates education more and more, students will become more aware and more expectant about it, asking for more gamified courses and curricula. In the next decade or so this transformation will become a necessity, so educators should also get ready for this change. If we are going to use serious games in our courses as well, as educators, we should learn more about game design so that we can design our own serious games for our own courses without the need to get outside help. It is obvious that the teacher of the course knows most about how to convey the knowledge that the students need to learn, hence designing a serious game to do just that should fall to the teacher as well.

With more data coming in from this year and years to come, students' approach to gamified in class applications coupled with a serious game project will become more and more clear. While these results are expected to follow last year's trend updates to the game will be made according to the feedback gained from the students.

REFERENCES

Abt, C. C. (1987). *Serious Games*. University Press of America.

Arai, S., Sakamoto, K., & Washizaki, H. (2014). *A Gamified Tool for Motivating Developers to Remove Warnings of Bug Pattern Tools.* Paper presented at the IWESEP 2014, Osaka, Japan. 10.1109/IWESEP.2014.17

Arnab, S., Berta, R., Earp, J., de Freitas, S., Popescu, M., Romero, M., . . . Usart, M. (2012). Framing the Adoption of Serious Games in Formal Education. *Electronic Journal of e-Learning, 10*(2), 59-171. Available online at www.ejel.com

Barr, M. (2018). Student attitudes to games-based skills development: Learning from video games in higher education. *Computers in Human Behavior, 80*, 283–294. doi:10.1016/j.chb.2017.11.030

Bellotti, F., Berta, R., De Gloria, A., Lavagnino, E., Antonaci, A., Dagnino, F., ... Mayer, I. S. (2014). Serious games and the development of an entrepreneurial mindset in higher education engineering students. *Entertainment Computing, 5*(4), 357–366. doi:10.1016/j.entcom.2014.07.003

Braghirolli, L., Ribeiro, J., Weise, A., & Pizzolato, M. (2016). Benefits of educational games as an introductory activity in industrial engineering education. *Computers in Human Behavior, 58*, 315–324. doi:10.1016/j.chb.2015.12.063

Breuer, J., & Bente, G. (2010). Why So Serious? On the Relation of Serious Games and Learning. *Eludamos (Göttingen), 4*(1), 7–24.

Coller, B., & Scott, M. (2009). Effectiveness of using a video game to teach a course in mechanical engineering. *Computers & Education, 53*(3), 900–912. doi:10.1016/j.compedu.2009.05.012

Crookall, D. (2010). Serious Games, Debriefing, and Simulation/Gaming as a Discipline. *Simulation & Gaming, 41*(6), 898–920. doi:10.1177/1046878110390784

Deif, A. (2017). Insights on lean gamification for higher education. *International Journal of Lean Six Sigma, 8*(3), 359–376. doi:10.1108/IJLSS-04-2016-0017

Deterding, S., Dixon, D., Khaled, R., & Nacke, L. (2011). *From Game Design Elements to Gamefulness: Defining "Gamification"*. Paper presented at the 15th International Academic MindTrek Conference, Tampere. 10.1145/2181037.2181040

Dias, J. (2017). Teaching operations research to undergraduate management students: The role of gamification. *International Journal of Management Education, 15*(1), 98–111. doi:10.1016/j.ijme.2017.01.002

Dijk, T. V., Spil, T., Burg, S. V., Wenzler, I., & Dalmolen, S. (2015). Present or Play, Some Evidence on the Effect on Behaviour of Serious Gaming. *International Journal of Game-Based Learning*, *5*(2), 55–69. doi:10.4018/ijgbl.2015040104

Dobrescu, L., Greiner, B., & Motta, A. (2015). Learning economics concepts through game-play: An experiment. *International Journal of Educational Research, 69*, 23–37. doi:10.1016/j.ijer.2014.08.005

Downes-Le Guin, T., Baker, R., Mechling, J., & Ruyle, E. (2012). Myths and realities of Driskell, J., & Dwyer, D. (1984). Microcomputer Videogame Based Training. *Educational Technology*, *24*(2), 11–16. Retrieved from http://www.jstor.org/stable/44427307

Entertainment Software Association. (2015). *2015: Sales, Demographic and Usage Data - essential Facts about the Computer and Video Game Industry*. Retrieved from http://www.theesa.com/wp-content/uploads/2015/04/ESA-Essential-Facts-2015.pdf

Everson, K. (2015). Learning Is All in the Wrist. *Chief Learning Officer*, *14*(4), 18–21.

Farzan, R., & Brusilovsky, P. (2011). Encouraging user participation in a course recommender system: An impact on user behavior. *Computers in Human Behavior*, *27*(1), 276–284. doi:10.1016/j.chb.2010.08.005

Fernandes, J., Duarte, D., Ribeiro, C., Farinha, C., Pereira, J. M., & Silva, M. M. (2012). iThink: A game-based approach towards improving collaboration and participation in requirement elicitation. *Procedia Computer Science, 15*, 66-77. doi:10.1016/j.procs.2012.10.059

Garris, R., Ahlers, R., & Driskell, J. E. (2002). Games, motivation, and learning: A research and practice model. *Simulation & Gaming*, *33*(4), 441–467. doi:10.1177/1046878102238607

Garris, R., Ahlers, R., & Driskell, J. E. (2002). Games, Motivation, and Learning: A Research and Practice Model. *Simulation & Gaming*, *33*(4), 441–467. doi:10.1177/1046878102238607

Gee, J. P. (2007). *Good video games and good learning: Collected essays on video games, learning, and literacy*. New York: Peter Lang International Academic Publishers. doi:10.3726/978-1-4539-1162-4

Gustafsson, A., Katzeff, C., & Bang, M. (2009). Evaluation of a pervasive game for domestic energy engagement among teenagers. Computers in Entertainment. *CIE*, *7*(4), 54. doi:10.1145/1658866.1658873

Hamari, J. (2013). Transforming homo economicus into homo ludens: A field experiment on gamification in a utilitarian peer-to-peer trading service. Electronic Commerce Research and Applications. *Social Commerce*, *12*(Part 2), 236–245. doi:10.1016/j.elerap.2013.01.004

Hamari, J. (2015). Do badges increase user activity? A field experiment on the effects of gamification. *Computers in Human Behavior*. doi:10.1016/j.chb.2015.03.036

Hauge, J. M. B., Pourabdollahian, B., & Riedel, J. C. K. H. (2013). The Use of Serious Games in the Education of Engineers. In C. Emmanouilidis, M. Taisch, & D. Kiritsis (Eds.), *Advances in Production Management Systems. Competitive Manufacturing for Innovative Products and Services. APMS 2012. IFIP Advances in Information and Communication Technology* (Vol. 397). Berlin: Springer. doi:10.1007/978-3-642-40352-1_78

Hense, J., & Mandl, H. (2014). Learning in or with games? Quality criteria for digital learning games from the perspectives of learning, emotion, and motivation theory. In D. G. Sampson, D. Ifenthaler, J. M. Spector, & P. Isaias (Eds.), Digital systems for open access to formal and informal learning (pp. 181-193). Piräus: Springer. doi:10.1007/978-3-319-02264-2_12

Ipsos. (2011). *Engagement Unleashed: Gamification for Business, Brands and Loyalty*. http://saatchi.com/en-us/news/engagement_unleashed_gamification_for_business_brands_and_loyalty

Jones, B. A., Madden, G. J., & Wengreen, H. J. (2014). The FIT game: Preliminary evaluation of a gamification approach to increasing fruit and vegetable consumption in school. *Preventive Medicine, 68*, 76-79. doi:10.1016/j.ypmed.2014.04.015

Kapp, K. M. (2012). *The gamification of learning and instruction: Game-based methods and strategies for training and education*. San Francisco: Pfeiffer.

Kasurinen, J., & Knutas, A. (2018). Publication trends in gamification: A systematic mapping study. *Computer Science Review*, *27*, 33–44. doi:10.1016/j.cosrev.2017.10.003

Kim, B. (2012). Harnessing the power of game dynamics1: Why, how to, and how not to gamify the library experience. *College & Research Libraries News*, *73*(8), 465–469. doi:10.5860/crln.73.8.8811

Kumar, J. (2013). Gamification at Work: Designing Engaging Business Software. Design, User Experience, and Usability. Health, Learning, Playing, Cultural, and Cross-Cultural User Experience. *Lecture Notes in Computer Science*, *8013*, 528–537. doi:10.1007/978-3-642-39241-2_58

Landers, R. N., & Landers, A. K. (2014). An Empirical Test of the Theory of Gamified Learning: The Effect of Leaderboards on Time-on-Task and Academic Performance. *Simulation & Gaming, 45*(6), 769-785. doi:10.1177/1046878114563662

Lee, J., Luchini, K., Michael, B., Norris, C., & Soloway, E. (2004, April). More than just fun and games: Assessing the value of educational video games in the classroom. In *CHI'04 extended abstracts on Human factors in computing systems* (pp. 1375–1378). ACM. doi:10.1145/985921.986068

Lee, J. J., & Hammer, J. (2011). Gamification in education: What, how, why bother? *Academic Exchange Quarterly*, *15*(2), 146.

Liu, Y., Alexandrova, T., & Nakajima, T. (2011). Gamifying intelligent environments. In *Proceedings of the 2011 international ACM workshop on ubiquitous meta user interfaces* (pp. 7-12). ACM. 10.1145/2072652.2072655

Locke, E. A., & Latham, G. P. (1990). *A theory of goal setting and task performance*. Englewood Cliffs, NJ: Prentice Hall.

Malliarakis, C., Tomos, F., Shabalina, O., Mozelius, P., & Balan, O. (2015). How to Build an Ineffective Serious Game: Worst Practices in Serious Game Design. *ECGBL*.

Malone, T. W., & Lepper, M. R. (1987). Making learning fun: A taxonomy of intrinsic motivations for learning. In R. E. Snow & M. J. Farr (Eds.), Aptitude, learning, and instruction: Vol. 3. Conative and affective process analyses (pp. 223-253). Hillsdale, NJ: Lawrence Erlbaum.

Markopoulos, A. P., Fragkou, A., Kasidiaris, P. D., & Davim, J. P. (2015). Gamification in engineering education and professional training. *International Journal of Mechanical Engineering Education*, *43*(2), 118–131. doi:10.1177/0306419015591324

Mekler, E. D., Brühlmann, F., Tuch, A. N., & Opwis, K. (2015). Towards understanding the effects of individual gamification elements on intrinsic motivation and performance. *Computers in Human Behavior*. doi:10.1016/j.chb.2015.08.048

Müller, B. C., Reise, C., & Seliger, G. (2015). Gamification in factory management education. *Procedia CIRP*, *26*, 121–126. doi:10.1016/j.procir.2014.07.056

Orlanda, Ramb, Langc, Houserd, Klinge, & Coccia. (2014). Saving energy in an office environment: A serious game intervention. *Energy and Buildings, 74,* 43-52. doi:10.1016/j.enbuild.2014.01.036

Pasin, F., & Giroux, H. (2011). The impact of a simulation game on operations management education. *Computers & Education*, *57*(1), 1240–1254. doi:10.1016/j.compedu.2010.12.006

Przybylski, A. K., Rigby, C. S., & Ryan, R. M. (2010). A motivational model of video game engagement. *Review of General Psychology*, *14*(2), 154–166. doi:10.1037/a0019440

Qian, M., & Clark, K. (2016). Game-based Learning and 21st century skills: A review of recent research. *Computers in Human Behavior*, *63*, 50–58. doi:10.1016/j.chb.2016.05.023

Rieber, L. P. (1996). Seriously considering play: Designing interactive learning environments based on the blending of microworlds, simulations, and games. *Educational Technology Research and Development*, *44*(2), 43–58. doi:10.1007/BF02300540

Rigby, C. S., & Ryan, R. M. (2011). *Glued to games: How video games draw us in and hold us spellbound.* Santa Barbara: Praeger.

Ryan, R. M., Rigby, C. S., & Przybylski, A. K. (2006). The motivational pull of video games: A self-determination theory approach. *Motivation and Emotion*, *30*(4), 344–360. doi:10.100711031-006-9051-8

Sailer, M., Hense, J. U., Mayr, S. K., & Mandl, H. (2017). Jan Ulrich Hense, Sarah Katharina Mayr, Heinz Mandl, 2017, How gamification motivates: An experimental study of the effects of specific game design elements on psychological need satisfaction. *Computers in Human Behavior*, *69*, 371–380. doi:10.1016/j.chb.2016.12.033

Sharma, R. R. (2017). A Competency Model for Management Education for Sustainability, 2017. *Vision (Basel)*, *21*(2), 10–15.

Shi, L., Cristea, A. I., Hadzidedic, S., & Dervishalidovic, N. (2014). Contextual gamification of social interaction towards increasing motivation in social E-learning. In *Advances in web-based learning-ICWL 2014* (pp. 116–122). Springer. doi:10.1007/978-3-319-09635-3_12

Smith, S., & Chan, S. (2017). Collaborative and Competitive Video Games for Teaching Computing in Higher Education. *Journal of Science Education and Technology*, *26*(4), 438–457. doi:10.100710956-017-9690-4

Subhash, S., & Cudney, E. (2018). Gamified learning in higher education: A systematic review of the literature. *Computers in Human Behavior*, *87*, 192–206. doi:10.1016/j.chb.2018.05.028

Werbach, K., & Hunter, D. (2012). *For the win: How game thinking can revolutionize your business*. Philadelphia: Wharton Digital Press.

Wiggins, B. E. (2016). An overview and study on the use of games, simulations, and gamification in higher education. *International Journal of Game-Based Learning*, *6*(1), 18–29. doi:10.4018/IJGBL.2016010102

Wilkinson, P. (2016). A Brief History of Serious Games. Entertainment Computing and Serious Games Lecture Notes in Computer Science, 17-41. doi:10.1007/978-3-319-46152-6_2

Yee, N. (2006). Motivations for play in online games. *Cyberpsychology & Behavior*, *9*(6), 772–777. doi:10.1089/cpb.2006.9.772 PMID:17201605

Zichermann, G., & Cunningham, C. (2011). Gamification by design: Implementing game machanics in web and mobile apps. Cambridge.

Chapter 11
Content Suggestion for Mobile Applications to Facilitate Student Life in Technical Universities:
The ITU Mobile

Cahit Ali Bayraktar
Istanbul Technical University, Turkey

ABSTRACT

All sorts of products, technologies, and devices are obsoleted more quickly than ever nowadays. Special, mobile applications, which make human life easier in a lot of ways, should always be up to date to avoid falling into disfavor. Thus, they need to keep pace with customer expectations. This study aims to present a way to offer content suggestions for university mobile applications. For this purpose, two focus groups on separate campuses of Istanbul Technical University (ITU) were formed in order to determine student expectations on the university's official mobile application (ITU Mobile). In addition to the 14 currently available features of the ITU Mobile, 32 additional features were identified as expectations, and the analytic hierarchy process (AHP) was employed to prioritize them. Results indicate that when designing mobile applications, universities should pay attention to several dimensions such as refreshment, life, education, facility, integration with other applications, and transportation to better facilitate their students' campus life.

DOI: 10.4018/978-1-7998-2562-3.ch011

INTRODUCTION

One of the sectors where the developments in technology can be observed in the easiest way is communication technologies. Products, new technologies, devices and applications, nowadays, are considered obsolete in a shorter period of time compared to the 20st century. The mobile phone sector is an innovation-based sector that uses advanced technology. Under this competitive structure, the market share of brands and their position in the market is constantly changing (Chan et al., 2006; Yıldız and Kurtuldu, 2013).

While the core benefit of mobile phones is expressed as mobile communication (calls, sms, etc.), it now has a use that is far beyond its core benefit such as socialization, information gathering, purchasing, and so on. Thanks to the advanced features of new mobile phones, different communication types such as voice messaging, text messaging, and multimedia have become easily portable from one point to another with technologies such as telephone signals, wireless connection, bluetooth and infrared. These developments have radically changed the structure of social networks, have led to a transformation in individual and social life (Mackenzie, 2006) and are increasingly playing an important role in our lives.

Mobile applications developed especially for Android and iOS operating systems play a major role in making smart phones multifunctional. According to Chun (2013), mobile phone applications are customer information systems that serve the purpose that consumers want to achieve, solve their problems, entertain and help them meet their needs. The process of developing these applications is based on the purposes and needs of consumers. Mobile applications can be used for professional purposes as well as for entertainment and communication purposes. According to Paoli (2012), users download gaming applications such as cards, puzzles, adventure and sports for entertainment purposes. They can also download functional applications such as calendars, calculators, notes, videos, audio and even banking applications.

Individuals, businesses, public institutions, military, educational institutions, non-governmental organizations along with many others. are developing mobile applications for different purposes (entertainment, education, alternative solutions to daily problems, convenience, efficiency, etc.) and some applications have global awareness and widespread use.

According to Chun, Chung and Shin (2013), a good mobile application should meet a requirement (if it does not meet the desired requirement, how appealing it seems does not matter), work efficiently (serve the purpose correctly for what it was made for), look attractive (users may consider non-appealing applications as poor quality), should focus on solving the current problem (should be free from unnecessary features that are not used), should be easy to apply (the application should be simple to use and install). Huy and Thanh (2012) listed the features that users expect from applications as ease of use, convenience, performance, functionality, and ability to work offline.

The aim of this study is to examine the expectations of Istanbul Technical University students from the ITU Mobile application and to develop development suggestions for it. Focus group studies were carried out with students to determine their intended use and suggestions for improvement. Analytic Hierarchy Process (AHP) technique was employed to sort and prioritize the suggestions.

Mobile Applications in Universities

According to Klein (2012), "A mobile application is a software application running on the operating system of a particular mobile device and downloaded to the device to perform a specific set of functions". McWherter and Gowell (2012) state that, although the decision to create a mobile application is difficult, a mobile application helps create brand awareness and improves interaction with customers (p. 17).

The combination of wireless technology and mobile computing results in increased transformations in the world of education (Alexander, 2004). Smartphones and mobile applications have become a daily essential requirement in the lives of higher education students (Green, Cantu, Wardle; 2014, Gupta, Gop, Kyei-Blankson, 2014).

In their study, Zhu, Marquez and Yoo (2015) focused on mobile applications related to the multimedia learning of university students, taking advantage of the rapid developments in educational technology. In particular, they developed a mobile application for Engineering Economics and made student satisfaction tests. Mishra, Jha, Umrah (2017) have studied the use of mobile application technology in library services, have examined the types of mobile applications, the benefits and potential drawbacks of mobile applications.

Campus Information System for Students is defined as "An interrelated group of information resources, accessible by computer through the campus institutional external and internal web environment, that a university places at the disposal of its users to enable them to consult it and/or provide a selection of significant and relevant data, in the wide context of their university life in its academic, administrative and social senses, in order to improve student's knowledge base" (Asif and Krogstie, 2011). In this research, Mobile Student Information System (MSIS) was defined in order to provide more user-centric information services to students. In this system, a service that integrates location information and different information services was proposed for students in the university. A questionnaire based on the use of older versions of the system and feedback collected from the students were proposed to develop the current solution.

As the digital generation continues to multiply in higher education institutions, universities will need to focus on being mobile. A survey conducted by Johnson, Means and Kay (2013) at the University of Florida in 2010-2011 showed that universities offers a variety of services that students, faculties and staff want to access from mobile devices to tuition, course fees, news about campus. As a result, mobile access to services is increasingly likely to become a decision point to help prospective students and staff decide where to join, where to work, and how to rank institutions. Therefore, it is becoming more and more important for institutions to be mobile (p. 11).

With the emergence of a growing wave of students using a variety of web-capable mobile devices and smartphones, universities need to develop a mobile presence. Research shows that students and faculty members and staff use mobile devices to a great extent (Madden, Lenhart, Duggan, Cortesi and Gasser, 2013; Amini et al., 2013; Johnson, Means, & Khey, 2013). Although there is much research to provide statistics and information on mobile use, there is little research on how universities should approach mobile communications and the development of mobile applications / mobile web (Valoris, 2015). Knowing the best practices available in mobile development and identifying whether existing universities apply the user experience approach to mobile development will help universities develop their mobile strategies.

Mobile applications have a very important place in campus life and make it easier to get better information about the opportunities offered by the university. Universities spend resources on developing mobile applications to improve student learning and campus life (Gupta, Gop, Kyei-Blankson, 2014). Green et al. stated that students are very interested in having an application and their primary preference is to access their courses and grades and to communicate with their professors (Green, Cantu, Wardle, 2014).

It is a digital campus concept that provides advanced networking and communication between university students by creating computer technology and integrated services. In the education sector, however, the digital campus needs to be built as a basis to modernize education and ensure a better campus life (Mahayuddin at al, 2018). It is envisaged that Mobile Applications, which are part of the Digital Campus concept, can contribute to the sustainability of campus. In addition, participation system, canteen payment and academic services can be made more effective with Mobile Applications. As the wireless, mobile learning experience evolves rapidly, universities should do their best to grasp the emerging trends.

MAIN FOCUS OF THE CHAPTER

Customer/Student Expectations

This study aims to collect and analyze the expectations of ITU Mobile application users as customers, and define the priorities of these criteria in order to develop this application according to customer expectations. The source of information of the study is the students of Istanbul Technical University as customers and ITU Mobile users.

Customer expectations are the emotions, opinions and the requirements that customers generate about a product or service. Customer expectations may depend on advertisements, experiences, word of mouth, and so on, and these expectations may be about quality, after sale services, price, and performance of the product.

Customers compare their experience with a product or service with their expectations, and they will be satisfied according to the degree of similarity between experience and expectation. In order to gain a competitive advantage through customer satisfaction, identifying and understanding customer expectations is crucial. Firms should develop their strategies, products and services according to the expectations of their customers. Being successful in identifying customer expectations and shape operations according to them leads satisfied customers; satisfied customers also lead repeated purchases, which lead customer loyalty.

Ojasalo (2001) stated that the nature of expectations can be fuzzy, precise, implicit, explicit, realistic and unrealistic. In some situations, customers expect an improvement or an alteration from a product or service; however, they are not able to clarify or make specifications about what they are expecting. These expectations are named as fuzzy expectations, whereas precise expectations are the total opposite of fuzzy expectations. Implicit expectations identify the expectations about an attribute, function, quality etc. that customers do not aware of expectations since they consider these attributes of a product or service already exist and the expectations become apparent when they are not met by the product or service. On the other hand, explicit expectations are the reverse of implicit expectations. These are the expectations such that customers are aware of what they want and if they feel unsatisfied from the product or service they can define the causes of dissatisfaction. Unrealistic expectations are the expectations that have really low possibility to be met by the firm, whereas realistic expectations are its opposite. Customer expectations can include all of these expectations at the same time; also, fuzzy, precise, explicit, implicit expectations can be realistic or unrealistic (Ojasalo, 2001).

According to Zeithaml et al. (1993), there are two levels of customer expectations. Each customer wants clarified benefits from products and services. The point that a customer expects to reach is defined as the desired level. However, desired expectations cannot be always met; hence, customers decrease the level of their expectations to a minimum level that they agree, which is called the adequate level. The gap between the desired level and the adequate level is called the zone of tolerance and the range of this zone can vary among different customers (Zeithaml et al., 1993).

It is possible to get information of customer expectations and turn it into the advantage of an organization by managing expectations well. In order to manage customer expectations, Parasuraman (1991) suggested that organizations should explain themselves to customers in order to prevent any inappropriate expectation that turns into unrealistic. Also, firms should avoid mistakes as much as possible because the increasing or non-reducing rates of mistakes damage the reliability of both the product or service and the firm, and it is crucial for a firm building trust of customers to manage expectations (Parasuraman et al., 1991).

ITU Mobile Application

ITU Mobile is an application that is designed by ITU Information Technologies Department, and it provides ITU members to use many functions in both university and campus life with their mobile phones. The main purpose of ITU Mobile is to facilitate as many functions that are used by ITU members as possible and to enable them to access information for different purposes. In order to do that, functions with different concepts and different purposes are integrated for the application. The features of ITU Mobile are listed in Table 1. Although the application offers a lot of functions for different aspects of university and campus life, it has been observed that students which are also considered as customers of the application, have feedbacks on the negative and positive sides of the application.

In the next section, methods for collecting and evaluating customer expectations are mentioned. This study includes two steps: the first step is for collecting expectations information and the second step is for evaluating expectations. Thus, the focus group study for collecting information and Analytic Hierarchy Process for analyzing collected information is selected to be implemented for the study.

As mentioned earlier, the focus group study is a method for searching information that enables researchers to access wide and deep information. Since it includes open-ended questions, the focus group study provides the ideas that can be unnoticed with structured methods. A customer expectations analysis study requires deeper information, and open-ended questions facilitates the process since the researchers cannot have complete information about the main subjects of information that will be gained through participants. Thus, in order to collect expectations for ITU Mobile application, two different focus group studies are planned to be conducted. One of the studies is planned to include participants who are currently the students of ITU on Ayazaga Campus, whereas the second study includes participants as ITU

Table 1. Features of ITU mobile

1. Campus maps, faculties and institutes in campus, social and cultural areas, sport facilities, food and drink places, information about other buildings and units
2. Help call to ITU health and safety unit in emergent situations with Emergency Button
3. The service times of shuttles and instantaneous location information of shuttles
4. Introduction tours of campus
5. Listening the Classic, Jazz/Blues and Rock stations of ITU Radio
6. Access to ITU Library Data Base, reservation and borrowing processes by integration of library system and ITU Mobile
7. Access to personal classes, homework and announcements by integration with Ninova
8. Screening ITU Profile, ITU Card Arrear and ITU Cloud
9. Accessing current news and announcements
10. Accessing ITU Phone Book and speed call
11. Creating Help Ticket for questions and problems
12. Daily Cafeteria Menu
13. Webmail connection
14. Access to information of pharmacies around the campus

students on Macka Campus. The aim of the focus group study is to learn about the students' awareness of ITU Mobile and their usage, the functions they are using most frequently, the drawbacks, their satisfaction with the product and their expectations. Open-ended questions are planned to be developed for these concepts.

In the second step of the study, AHP methodology is employed. The main goal here is to select the expectations that will improve the product best, and the criteria of the AHP model is planned to be designed according to expectations. It is expected to obtain both criteria and subcriteria for the model since the functions of the product varies among different usage areas such as education, campus life, and facilities. Following the model designation, randomly selected participants are asked to be interviewed to define the importance of criteria and subcriteria under consideration.

Methods for Collecting Information of Customer/Student Expectations

Methods such as focus groups, deep interwievs, KANO model, surveys / questionnaires, Quality Function Deployment, Analytic Hierarchy Process (AHP), and nominal group techniques can be used to determine customer / student expectations. In this study, a focus group was used to mine suggestions for the ITU Mobile application and Analytic Hierarchy Process (AHP) was employed for the prioritization of these suggestions.

Focus Group

A focus group generally consists of 6 to 10 people who are brought together to discuss a particular topic. Participants are chosen consciously according to their relevancy with the topic and groups can be constituted according to age, education, gender, etc. The number of participants in a focus group should be enough to provide a discussion and also convenient to manage. In order to collect customer expectations, customers that are chosen as focus group participants are considered to be the representatives of a firm's target customers. Focus group studies are conducted by a researcher which can be named also as a moderator, and the researcher aims to gather information about customers' opinions, feelings, perceptions, expectations, etc. The moderator guides and monitors the interactions and discussions.

The questions asked are open-ended and the results are qualitative. They may be summaries, headings or major themes. Even though the method does not result in accurate findings such as metrics or numerical data, it provides deeper information about customers' opinions, and also it may help to enlighten the expectations that customers are not aware of. This can be attributed to the fact that even the questions of researchers have a logical order and follow-up questions may be asked based on the participants' responses.

Even though the focus group method provides wider information, it can be costly and difficult to manage. Also, due to interactions between participants, views can be influenced and results may be distorted. Zeithaml, Berry and Parasuraman (1993) conducted a focus group study with size 9 participants in order to analyze customer expectations.

Analytic Hierarchy Process (AHP)

In order to evaluate the priorities of customer expectations and determine the expectations that should be satisfied primarily, Analytic Hierarchy Process (AHP) methodology can be adopted. AHP is a multi-criteria decision making tool which is developed by Thomas Saaty in 1980. The method used in decision making processes including several decision makers and criteria; it also helps making a decision among several alternatives. AHP allows a hierarchical model that represents the relationship among the main goal of the problem, the criteria and the alternatives. The method works when determining priorities for several alternatives and the criteria in order to judge these alternatives (Saaty, 1990).

It is difficult for decision makers to give the best solution among several alternatives when more than one criterion is given for evaluation. However, when it is asked that which one of the two alternatives is better or which one of the two criteria is more important, giving an answer will be easier. Thus, AHP reduces a complex decision-making problem (multi-criteria and multi-alternatives) to pairwise comparisons for obtaining a solution.

The steps of AHP (Vaidya & Kumar, 2006);

Step 1: Determining the problem and criteria that has impact on decision

Step 2: Developing the structure of decision hierarchy

The problem that is solved with AHP requires as detailed definitions as possible. These definitions are determined by a hierarchy. AHP problems are defined at least on three levels; the level that is located at the top of the hierarchy includes the goal, and the lowest level usually includes the alternatives (Figure 1). The important point in AHP is that an element on a level is independent from all the other elements in the hierarchy.

Figure 1. A three level hierarchy (Saaty & Vargas, 2001)

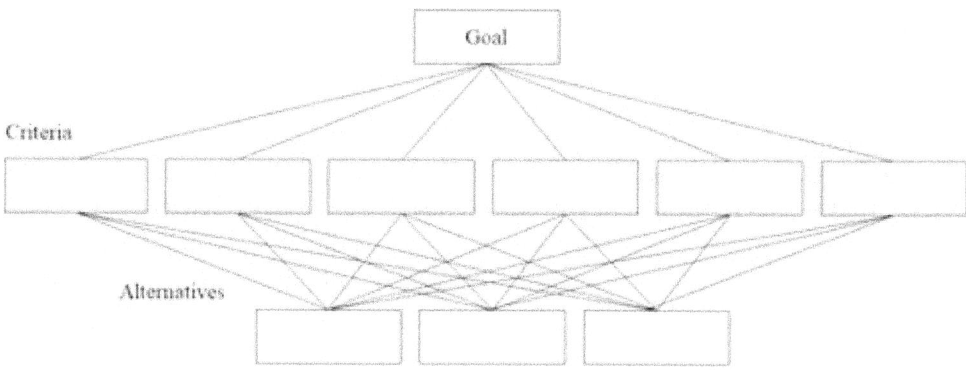

Step 3: Constructing the set of pairwise comparisons matrix

Pairwise comparisons include comparing two elements' priorities according to a predetermined scale. This scale is called as the Fundamental Scale by Saaty, and it shows how many times an element is more important or more dominant than another when comparing the element according to a comparison criterion. The number of comparisons is equal to $n(n-1)/2$, where n represents the number of elements. Table 2 below examplifies the fundamental scale of the pairwise comparison matrix for relative consumption of drinks.

Step 4: Calculation of priorities

Priorities are calculated from the normalized pairwise comparisons matrix and they are equal to the average of each row.

Table 2. The fundamental scale

Intensity of Importance	Definition	Explanation
1	Equal importance	Two activities contribute equally to the objective
2	Weak or slight	
3	Moderate importance	Experinece and judgement slightly favour one activity over another
4	Moderate plus	
5	Strong importance	Experinece and judgement strongly favour one activity over another
6	Strong plus	
7	Very strong or demonstrated importance	An activity is favoured very strongly over another; its dominance demonstrated in practice
8	Very, very strong	
9	Extreme importance	The evidence favouring ane activity over another is of the highest possible order of affirmation
Reciprocals of above	If activity *i* has one of the above non-zero numbers assigned to it when compared with activity *j*, then *j* has reciprocal value when compared with *i*	A reasonable assumption
1.1 – 1.9	If the activities are very close	May be difficult to assign the best value but when compare with other constrasting activities the size of small numbers would not to be too noticeable, yet they can still indicate the relative importance of the activities

Step 5: Synthesis and validation

Pairwise comparison matrices include comparisons of alternatives for each criterion separately, in this step, all of the criteria are synthesed to one matrix. For each column of the new matrix, priorities of each alternative according to each criterion is selected. Then, the compromised weights are calculated and it leads to the decision.

The validation step includes executing the calculations of the Eigen vector, the consistency index and the consistency ratio for each criterion. If the values are satisfactory, the decision is made; if they are not, the process will be repeated until the desired values are obtained.

Findings

Focus Group Studies

One of the focus group studies was conducted on ITU Ayazaga Campus with 10 participants, and the second one was conducted on ITU Macka Campus with the same number of participants. All of the participants were ITU students from a total of 9 departments; namely, aeronautical engineering, astronautical engineering, industrial engineering, management engineering, computer engineering, mechanical engineering, food engineering, environmental engineering, and geomatics engineering. The focus groups included freshman students, sophomores, juniors and seniors. 5 open-ended questions and 1 closed ended ranking question addressed are shown in Table 3.

Table 3. Focus group study questions

1. Do you use ITU Mobile?
2. Which features are you aware of? Which of them do you use more often?
3. Which features ITU Mobile lacks now may motivate you to use this application in the future?
4. How would you rank your satisfaction from 1 to 6? (1- I am completely dissatisfied, 2- I am dissatisfied, 3- I am a little bit dissatisfied, 4- I am a little bir satisfied, 5- I am satisfied, 6- I am completely satisfied)
5. Which functions do you use most frequently?
6. What functions would you wish to be included in this application?

All the participants are the users of ITU Mobile application and Ninova; access to daily cafeteria menu, access to information of shuttle hours and access to news and announcements are the functions that are mostly used by participants. Based on the answers, it is shown that the average customer satisfaction rating of ITU Mobile is 3.5/6. The drawbacks and possible reasons of low satisfaction observed can be stated as technical problems, interface problems, lack of integration with other applications. The nice to have features can also be listed as Ninova integration, design of interface, including a lot of features in one application, and accessibility. The customer expectations collected from the participants are shown in Table 4. Expectations #7, #8 and #9 are not considered in the evaluation since these functions can be executed with the current system, and expectation #28 is not considered either

Table 4. Student expectations for ITU mobile

Student Expectations	Group
1. Integration of Bizero bicycle application	Integration with other applications
2. Notifications of Webmail	Integration with other applications
3. Integration of ITU Portal website	Integration with other applications
4. Access to notifications, materials, information of preparatory and common courses	Education
5. Access to annoucements of mandatory internship programs	Education
6. Function for desinging course program with open courses	Education
7. Notifications of courses for starting	Education
8. Obtaining Ninova classes from sis.itu.web.tr	Education
9. Accessing notificated files of Ninova notifications with extension link	Education
10. Access to academical publishes, format informations for articles etc.	Education
11. Calendar for students to follow absences	Education
12. Viewing common lab course classes	Facility
13. Viewing ATM points	Facility
14. Viewing floor plans of department buildings	Facility
15. Reporting lost/stolen ITU cards	Facility
16. Creating request for student certificate, transkript from Student Affairs	Facility
17. Taxi hailing system	Transportation
18. ITU Navigation	Transportation
19. Access to information of shuttle stops and remaining time to arrive	Transportation
20. Access to information of public transportation that includes campuses	Transportation
21. Map of intensity for library	Life
22. Creating request of permition for dormitories	Life
23. Reservation system for facilities in the campus	Life
24. Make appointment for infirmary	Life
25. Accessing information of social clubs, getting notification of followed clubs	Life
26. Following events and getting notifications	Life
27. Connection to ARI24	Life
28. Button for Emergency Help	Life
29. Announcements of job opportunities in campus departments	Life
30. Money transfer from bankcard to ITU card for cafeteria usage	Refreshments
31. Online order from restaurants in campus	Refreshments
32. QR code for accessing cafeteria with ITU Mobile	Refreshments

Figure 2. AHP Model for student expectations of ITU mobile

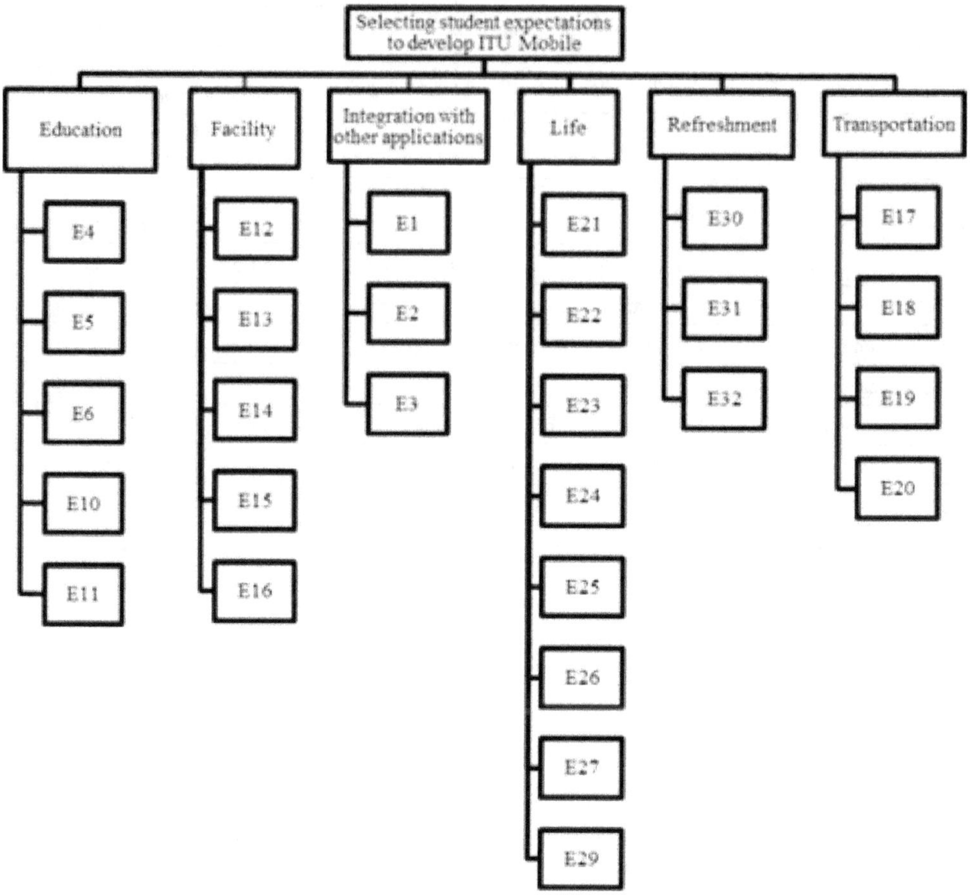

since it is already available in the application. Also, these expectations are grouped according to their usage areas as integration with other applications, education, facility, transportation, life and refreshments.

AHP Model

After collecting the customer expectations with focus group studies, an AHP model is designed to determine the expectations that are mostly preferred by students, which are considered as customers. The goal of the AHP model is defined as selecting the customer expectations to develop ITU Mobile. As the first step of AHP, the model is designed according to expectations in Table 4. Subcriteria are defined as customer expectations, and the criteria are defined as groups which are referred to as usage area of ITU Mobile application. The model is shown in Figure 2. After conducting the AHP model, pairwise comparisons of the criteria are executed by interviewing participants which are also students of ITU, 3 interviews are conducted in order to construct the pairwise comparisons matrix. Comparisons are made for both criteria and subcriteria in its own groups. In the comparisons, questions to defin the importance are asked to participants using The Fundamental Scale given in Table 2.

In this study, expectations other than the features of ITU Mobile Application were questioned. Therefore, during the AHP analysis, the existing features were excluded from the analysis and the focus was on determining other expectations.

Table 5. Pairwise comparison matrix for criteria

Criteria	Education	Facility	Integration with other applications	Life	Refreshment	Transportation	Priorities
Education	1.0000	1.0000	0.5000	1.0000	1.0000	3.0000	0.1592
Facility	1.0000	1.0000	1.0000	2.0000	0.2000	2.0000	0.1528
Integration with other applications	2.0000	1.0000	1.0000	0.5000	0.5000	1.0000	0.1273
Life	1.0000	0.5000	2.0000	1.0000	0.5000	3.0000	0.1698
Refreshment	1.0000	5.0000	2.0000	2.0000	1.0000	4.0000	0.3184
Transportation	0.3333	0.5000	1.0000	0.3333	0.2500	1.0000	0.0725
Consistency ratio 0.09 < 0.1							

Table 6. Pairwise comparison matrix for education criterion

Subcriteria	E4	E5	E6	E10	E11	Priorities
E4	1.0000	0.1667	0.1250	0.1250	0.1250	0.0298
E5	6.0000	1.0000	1.0000	0.5000	0.5000	0.1738
E6	8.0000	1.0000	1.0000	4.0000	0.5000	0.2800
E10	8.0000	2.0000	0.2500	1.0000	0.5000	0.2269
E11	8.0000	2.0000	2.0000	2.0000	1.0000	0.2896
Consistency ratio 0.09 < 0.1						

The compromised importance of pairwise comparisons is calculated with Super Decisions program and the tables are given below. The row weights which are obtained from the Eigen vectors of pairwise comparison matrices show the priorities of the criteria.

Table 7. Pairwise comparison matrix for facility criterion

Subcriteria	E12	E13	E14	E15	E16	Priorities
E12	1.0000	3.0000	0.5000	0.5000	0.2500	0.1103
E13	0.3333	1.0000	0.3333	0.2000	0.1250	0.0418
E14	2.0000	3.0000	1.0000	0.2000	0.1667	0.1337
E15	2.0000	5.0000	5.0000	1.0000	1.0000	0.2941
E16	4.0000	8.0000	6.0000	1.0000	1.0000	0.4201
Consistency ratio 0.065 < 0.1						

Table 8. Pairwise comparison matrix for transportation criterion

Subcriteria	E17	E18	E19	E20	Priorities
E17	1.0000	1.0000	0.1429	0.5000	0.0896
E18	1.0000	1.0000	0.1667	2.0000	0.1412
E19	7.0000	6.0000	1.0000	5.0000	0.6439
E20	2.0000	0.5000	0.2000	1.0000	0.1254
Consistency ratio 0.06 < 0.1					

According to the priority values in Table 5, the most important expectations of the students from ITU Mobile application are refreshment, life and education. On the other hand, it can be seen that transportation has a considerably small value. Thus, it can be suggested that product developments of ITU Mobile should be executed to refreshment, education and life criteria primarily.

For education criteria itself, it can be observed that calendar for students to follow absences (E11) and function for designing course program with open courses (E6) are the most important expectations that should be considered with priority values according to Table 6. Also, access to academic publications, format information for articles (E10) should also be considered in order to develop ITU Mobile.

Table 9. Pairwise comparison matrix for integ. with other app. criterion

Subcriteria	E1	E2	E3	Priorities
E1	1.0000	0.5000	0.1429	0.1085
E2	2.0000	1.0000	0.5000	0.2311
E3	7.0000	2.0000	1.0000	0.6604
Consistency ratio 0.068 < 0.1				

Table 10. Pairwise comparison matrix for refreshment criterion

Subcriteria	E30	E31	E32	Priorities
E30	1.0000	9.0000	2.0000	0.6057
E31	0.1111	1.0000	0.2000	0.0662
E32	0.5000	5.0000	1.0000	0.3281
Consistency ratio 0.01 < 0.1				

Creating request for student certificate, transcript from Student Affairs (E16); and reporting lost/stolen ITU cards (E15) are the expectations that should be considered firstly, the priority values can be seen in Table 7.

According to the priority values in Table 8, the most important customer expectation of transportation criteria is access to information of shuttle stops and remaining time to arrive (E19) with considerably large amount of value 0.65. Other expectations, ITU Navigation (E18), access to information of public transportation that includes campuses (E20) and taxi hailing system (E17) reppectively.

Students as customers of ITU Mobile prefer an integration with ITU Portal website (E3) for ITU Mobile application and Notifications of Webmail (Table 9) in the aspect of integration with other applications criteria.

On the other hand, money transfer from bankcard to ITU card for cafeteria usage (E30) is highly expected by students, and QR code for accessing cafeteria with ITU Mobile (E32) also has considerably high priority value for refreshment criteria (Table 10).

Lastly, it can be seen that life criteria have the largest number of expectations by students. From Table 11, Reservation system for facilities in the campus (E23) and following events and getting notifications (E26) are the most expected elements of these criteria. Also, accessing information of social clubs, getting notification of followed clubs (E25) and map of intensity for library (E21) should be considered for developing the application.

Table 11. Pairwise comparison matrix for life criterion

Subcriteria	E21	E22	E23	E24	E25	E26	E27	E29	Priorities
E21	1.0000	2.0000	0.5000	0.5000	1.0000	0.2500	7.0000	2.0000	0.1314
E22	0.5000	1.0000	0.5000	0.5000	0.5000	1.0000	2.0000	1.0000	0.0645
E23	2.0000	2.0000	1.0000	4.0000	3.0000	1.0000	8.0000	6.0000	0.2490
E24	2.0000	2.0000	0.2500	1.0000	0.5000	0.5000	6.0000	2.0000	0.1314
E25	1.0000	2.0000	0.3333	2.0000	1.0000	0.5000	7.0000	2.0000	0.1460
E26	4.0000	1.0000	1.0000	2.0000	2.0000	1.0000	7.0000	4.0000	0.2029
E27	0.1429	0.5000	0.1250	0.1429	0.1429	0.1429	1.0000	1.0000	0.0295
E29	0.5000	1.0000	0.1667	0.5000	0.5000	0.2500	1.0000	1.0000	0.0453
Consistency ratio 0.03 < 0.1									

SOLUTIONS AND RECOMMENDATIONS

Universities benefit from the rapid developments in mobile technologies both in the learning of university students and in facilitating university life. In the related literature, it is seen that many mobile applications focused on education and finding location on campus have been developed and these applications have been examined in terms of usability. Apart from these two basic expectations of the students, they need other information that will make their campus life easier, accelerate their adaptation to the university environment and make their lives easier. When these expectations are evaluated in terms of "Technical University" and facilitating campus life, it can be stated that they are gathered under five main dimensions. These are according to the order of AHP analysis: refreshment, life, education, facility, integration with other applications and transportation.

As the study focused on expectations other than the features of ITU Mobile Application, the following features were not evaluated in the study.

- Campus maps (faculties and institutes in campus, social and cultural areas, sport facilities, food and drink places, information about other buildings and units)
- Help call to ITU health and safety unit in emergent situations with Emergency Button
- The service times of shuttles and instantaneous location information of shuttles
- Introduction tours of campus
- Listening the Classic, Jazz/Blues and Rock stations of ITU Radio

- Access to ITU Library Data Base, reservation and borrowing processes by integration of library system and ITU Mobile
- Access to personal classes, homework and announcements by integration with Ninova
- Screening ITU Profile, ITU Card Arrear and ITU Cloud
- Accessing current news and announcements
- Accessing ITU Phone Book and speed call
- Creating Help Ticket for questions and problems
- Daily Cafeteria Menu
- Webmail connection
- Access to information of pharmacies around the campus

Developing integrated mobile applications that include these five main dimensions will facilitate the campus life of students. The features that should be found in the mobile applications developed by universities can be specified as follows:

Refreshment

Refreshment can be defined as facilitating the expenditures made on the university campus. Students considered as the most important factor to facilitate their university campus life as being able to use their University Identity Card or Mobile applications instead of using debit cards in their spending on campus. The ability to order from mobile applications in restaurants and cafeterias is another important factor.

Life

Life can be defined as the expectation to learn about social life and activities on the university campus. When the expectations under this main dimension are examined, the expectation of making an appointment through the mobile application from the facilities working with the appointment system in the campus stands out. Information about the activities of social and student clubs, reminder before the activities, access to library density map, being able to have information about job opportunities on campus are other features that should be included in this dimension.

Education

Education can be defined as the student's ability to follow course materials, internship information and documents, academic publications, absences. When preparing mobile applications, the features to be considered respectively can be defined as follows: Calendar for students to follow absences, Function for desinging course program,

Access to academical publishes, format informations for articles etc., Access to annoucements of mandatory internship programs, Access to notifications, materials, information of preparatory and common courses.

Facility

Facility can be defined as creating student certificates and transcripts, viewing laboratory classrooms, viewing building floor plans, access to campus plan, and loss/stolen notification. While creating their own mobile applications, universities can add the ability to create student documents and transcripts, make lost / stolen notifications, and access building floor plans through the mobile application. These features emerge as the most important expectations that facilitate students' campus life in this dimension.

Integration with Other Applications

In addition to the university's own mobile applications, there may be different mobile applications of the student relations unit, administrative units (information technology, health-culture-sports, etc.), email software (webmail) and operating private companies. The university's mobile application should be integrated with other applications to facilitate student campus life. In particular, it is recommended that the mobile applications of the administrative units and the email application be integrated with the general mobile application of the university.

Transportation

Transportation can be defined as the expectation to receive continuous information about transportation within the campus. Under this dimension, it is recommended that a navigation system covering the stops and transportation times of the university's shuttle vehicles and public transportation on campus should be included in the university's mobile application.

FUTURE RESEARCH DIRECTIONS

The present study investigating the characteristics of mobile applications launched by Technical Universities can be improved by diversifying the research methods used in determining the expectations.

Here, the expectations of students on the mobile application of technical universities were determined with the focus group method. The recommendations formed as a result of two focus group studies were prioritized with AHP analysis. Today, the use of business analysis methods in the development of mobile applications is becoming a more and more applied technique. In addition to Business Analysis Techniques, KANO Model, and Quality Function Deployment (QFD) techniques can be used in the analysis and prioritization of student expectations.

Business analysis is the practice of enabling change in an enterprise by defining needs and recommending solutions that deliver value to stakeholders. Business analysis enables an enterprise to articulate needs and the rationale for change, and to design and describe solutions that can deliver value. Business analysis can be performed from a diverse array of perspectives: agile, business intelligence, information technology, business architecture, and business process management. By analyzing the 50 analysis techniques defined in the BABOK Guide (The Business Analysis Body of Knowledge), other techniques suitable for the research subject can be used (IIBA, 2015).

Kano model is developed by Noriaki Kano and his associates in 1984, and it is used to categorize customer needs and expectations. The model classifies product features as customer expectations, and it claims that there exists a relationship between the degree of meeting customer expectations and customer satisfaction. According to this model, the requirements of products that are expected by customers can be classified into three main categories and each of these affects customer satisfaction differently (Kano et al., 1984); Must-be requirements, One-dimensional requirements, Attractive requirement.

Quality Function Deployment (QFD) is developed by Yoji Jakao in Japan in 1966 and firstly be adapted to Mitsubishi Heavy Industries Kobe Shipyard (Sullivan, 1996). "QFD is a method for developing a design quality aimed at satisfying the consumer and then translating customer's demand into design targets and major quality assurance points to be used throughout the production phase..." (Akao, 1990). In other words, QFD is an approach that is used commonly for product development and it identifies customer expectations and uses these expectations to develop plans for production in order to meet customer expectations. QFD aims to be able to prioritizing customer needs, translating these needs into technical characteristics of products, and satisfying customers.

CONCLUSION

When smart phones, which became the basic requirement of higher education students, are integrated with the rapid developments in educational technology, not only mobile applications related to education but also mobile applications that facilitate university campus life have started to develop. In mobile applications launched, services are provided in many areas such as campus map, access to course contents, library services, student services, etc. Students need more content than these services to make campus life easier.

It can be stated that these expectations are gathered under five main dimensions. The students also stated their sub-expectations in terms of each expectation dimension. Universities, which develop their own mobile applications, should give importance to the dimensions of refreshment, life, education, facility, integration with other mobile applications and transportation in order to facilitate the campus life of the students.

It is determined that the currently used ITU Mobile application does not fully meet the expectations of the students. For this reason, a mobile application content is proposed that will meet the basic expectations of Technical University students as listed below:

- Facilitates expenditures and payments made on the university campus,
- Provides quick access to information on social life and activities on the university campus,
- Provide fast and continuous access to course materials, internship information and documents, academic publications,
- Enables students to generate student certificates and transcripts and view laboratory classrooms and building and floor plans.
- Integrated with different mobile applications of administrative units (student affairs, information technology, health-culture-sports), e-mail and mobile applications of private companies,
- Provides fast and continuous access to on-campus transportation information.

The realization of the suggestions given above will make the campus life of students very easy in the digital age. Thus, students will be able to spend their time on their education, personal development and social life.

REFERENCES

Akao, Y. (1990). *Quality Function Deployment*. Cambridge, MA: Productivity Press.

Alexander, B. (2004). Going Nomadic: Moblile Learning in Higher Education. *EDUCAUSE Review, 39*(5).

Amini, M., Blair, M. A., Forrester, J., Goldstein, S. J., Katsouros, M., Rocchio, R. A., & Williams, A., III. (2013). *Developing a campus mobile strategy: Guidelines, tools, and best practices*. Retrieved from https://library.educause.edu/-/media/files/library/2013/1/acti1303-pdf.pdf

Asif, M., & Krogstie, J. (2011). Mobile student information system. *Campus-Wide Information Systems*, *28*(1), 5–15. doi:10.1108/10650741111097269

Chan, J., Chen, Z., Cormane, I., Her, N., & Thomas, R. (2006). *Cell Phone Industry Analysis Report*. Sacremanto State University Press.

Chun, S. G., Chung, D., & Shin, Y. (2013). Are Students Satisfied with The Use of Smartphone Apps. *Issues in Information Systems*, *14*(2), 23–33.

Green, M., Cantu, A., & Wardle, A. (2014). A study examining student preferences of mobile applications at an institution of higher learning in South Texas. In T. Bastiaens (Ed.), *Proceedings of World Conference on E-Learning* (pp. 717-723). New Orleans, LA: Association for the Advancement of Computing in Education (AACE).

Gupta, P., Gop, K., & Kyei-Blankson, L. (2014). College students' usage of and expectations from university owned mobile applications. In T. Bastiaens (Ed.), *Proceedings of World Conference on E-Learning* (pp. 742-745). New Orleans, LA: Association for the Advancement of Computing in Education (AACE).

Huy, N. P., & Van Thanh, D. (2012). Evaluation of Mobile App Paradigms. *10th International Conference on Advances in Mobile Computing & Multimedia*, 25-30.

IIBA, (2015). *A Guide to the Business Analysis Body of Knowledge (BABOK Guide)*. International Institute of Business Analysis (IIBA).

Johnson, D., Means, T., & Khey, D. N. (2013). *A state of flux: Results of a mobile device surveyat the University of Florida*. EDUCAUSE.

Kano, N., Seraku, N., Takahashi, F., & Tsuji, S. (1984). Attractive quality and Must-be quality. *The Journal of the Japanese Society for Quality Control*, *14*(2), 39–48.

Klein, D. (2012, Feb.). How to decide: Mobile websites vs. mobile apps. *Inspire*.

Mackenzie, D. L. (2006). *What Works in Corrections*. Cambridge: Cambridge University Press. doi:10.1017/CBO9780511499470

Madden, M., Lenhart, A., Maeve, D., Sandra, C., & Gasser, U. (2013). *Teens and Technology 2013*. Washington, DC: Pew Research Center.

Mahayuddin, Z.R., Suwadi, N.A., Jenal, R., & Arshad, H. (2018). T. Implementing smart mobile application to achieve a sustainable campus. *International Journal of Supply Chain Management*, *7*(3), 154–159.

McWherter, J., & Gowell, S. (2012). *Professional mobile application development*. Indianapolis, IN: John Wiley & Sons, Inc.

Mishra, A. S., Jha, J. K., & Umre, S. K. (2017). Mobile app and the library services. *International Journal of Information Library and Society*, *6*(2), 27–32.

Ojasalo, J. (2001). Managing customer expectations in professional services. *Managing Service Quality*, *11*(3), 200–212. doi:10.1108/09604520110391379

Paoli, C. (2012). Google, Microsoft, & Apple Agree on Mobile Privacy Accord. *ADTMag*. https://adtmag.com/articles/2012/02/23/mobile-privacy-accord.aspx

Parasuraman, A., Zeithaml, V. A., & Berry, L. L. (1991). SERVQUAL: A Multiple-Item Scale for Measuring Consumer Perceptions of Service Quality. *Journal of Retailing*, *64*(1), 12–40.

Saaty, T. L. (1990). How to make a decision: The Analytic Hierarchy Process. *European Journal of Operational Research*, *48*(1), 9–26. doi:10.1016/0377-2217(90)90057-I

Saaty, T. L., & Vargas, L. G. (2001). *Models, Methods, Concepts and Applications of the Analytic Hierarchy Process*. Norwell: Kluwer Academic Publishers. doi:10.1007/978-1-4615-1665-1

Sullivan, L. P. (1986). Quality function deployment. *Quality Progress*, *19*, 39–50.

Vaidya, S., & Kumar, S. (2006). Analytic hierarchy process: An overview of applications. *European Journal of Operational Research*, *169*(1), 1–29. doi:10.1016/j.ejor.2004.04.028

Valoris, M. (2015). A mixed-methods analysiis of best practices for land-grant unıversity mobile applications from a user experience design perspective (Master's thesis). Colorado State University. Libraries.

Yıldız, S., & Kurtuldu, H. S. (2013). Factors Affecting Electronic Service Brand Equity. In *Transcultural Marketing for Incremental and Radical Innovation* (pp. 434–492). IGI Global.

Zeithaml, V. A., Berry, L. L., & Parasuraman, A. (1993). The Nature and Determinants of Customer Expectations of Service. *Journal of the Academy of Marketing Science*, *21*(1), 1–1. doi:10.1177/0092070393211001

Zhu, W., Marquez, A., & Yoo, J. (2015). Engineering economics jeopardy! Mobile app for university students. *The Engineering Economist*, *60*(4), 291–306. doi:10.1080/0013791X.2015.1067343

ADDITIONAL READING

Amini, M., Blair, M. A., Forrester, J., Goldstein, S. J., Katsouros, M., Rocchio, R. A., Williams, A., III. (2013). *Developing a campus mobile strategy: Guidelines, tools, and best practices.*

Green, M., Cantu, A., & Wardle, A. (2014). A study examining student preferences of mobile applications at an institution of higher learning in South Texas. In T. Bastiaens (Ed.), *Proceedings of World Conference on E-Learning* (pp. 717-723). New Orleans, LA, USA: Association for the Advancement of Computing in Education (AACE).

IIBA, (2015). *A Guide to the Business Analysis Body of Knowledge (BABOK Guide)*, International Institute of Business Analysis (IIBA).

Kano, N., Seraku, N., Takahashi, F., & Tsuji, S. (1984). Attractive quality and Must-be quality. *The Journal of the Japanese Society for Quality Control, 14*(2), 39–48.

Liu, Y., Shou, G., Hu, Y., …. (2017), *Towards a smart campus: Innovative applications with WiCloud platform based on mobile edge computing*, 12th International Conference on Computer Science and Education (ICCSE), IEEE (2017), pp. 133-138 10.1109/ICCSE.2017.8085477

Mahayuddin, Z.R., Suwadi, N.A., Jenal, R., & Arshad, H., Adiono. (2018). T. Implementing smart mobile application to achieve a sustainable campus. *International Journal of Supply Chain Management, 7*(3), 154–159.

Sullivan, L. P. (1986). Quality function deployment. *Quality Progress, 19*, 39–50.

Valoris, M. (2015), A mixed-methods analysıis of best practices for land-grant unıversity mobile applications from a user experience design perspective, A. Masters Theses, Colorado State University. Libraries

KEY TERMS AND DEFINITIONS

Focus Group: A focus group is a gathering of deliberately selected people who participate in a planned discussion intended to elicit consumer perceptions about a particular topic or area of interest in an environment that is non-threatening and receptive. Focus groups are a collective on purpose. Unlike interviews, which usually occurs with an individual, the focus groups allow members of a group to interact and influence each other during the discussion and consideration of ideas and perspectives.

Mobil Applications: A mobile application, also referred to as a mobile app or simply an app, is a computer program or software application designed to run on a mobile device such as a phone, tablet, or watch.

Student/Customer Expectations: Customer expectation encompasses everything that a customer expects from a product, service or organisation. Customer expectations are created in the minds of customers based upon their individual experiences and what they have learned, combined with their pre-existing experience and knowledge.

Student/Customer Satisfaction: Student/customer satisfaction is defined as a measurement that determines how happy students/customers are with a company's products, services, and capabilities. Student/customer satisfaction information, including surveys and ratings, can help a company determine how to best improve or changes its products and services.

Compilation of References

ABET. (2019). *ABET accreditation criteria for accrediting engineering programs in 2019-2020.* Retrieved from https://www.abet.org/accreditation/accreditation-criteria/criteria-for-accrediting-engineering-programs-2019-2020/]

ABET. (2019). *ABET accredited program details.* Retrieved from http://main.abet.org/aps/AccreditedProgramsDetails.aspx?OrganizationID=554&ProgramIDs=

ABET. (2019). *ABET history*. Retrieved from https://www.abet.org/about-abet/history/

ABET. (2019). *Accreditation changes*. Retrieved from https://www.abet.org/accreditation/accreditation-criteria/accreditation-changes/

Abowd, G. D. (n.d.). Beyond Weiser: From Ubiquitous to Collective. Computing. *IEEE Computer, 49*(1), 17–23.

Abt, C. C. (1987). *Serious Games*. University Press of America.

Adams, D. M., Mayer, R. E., MacNamara, A., Koenig, A., & Wainess, R. (2012). Narrative games for learning: Testing the discovery and narrative hypotheses. *Journal of Educational Psychology, 104*(1), 235–249. doi:10.1037/a0025595

Ajjan, H., & Hartshorne, R. (2008). Investigating faculty decisions to adopt Web 2.0 technologies: Theory and empirical tests. *The Internet and Higher Education, 11*(2), 71–80. doi:10.1016/j.iheduc.2008.05.002

Akao, Y. (1990). *Quality Function Deployment*. Cambridge, MA: Productivity Press.

Akdemir, D. A. (2007). *Yönetici Değerlerinin, Eğitim Örgütleri Karar Sürecine Etkileri* (Unpublished master's thesis). Fırat Üniversitesi, Elazığ.

Al Musawi, A. (2011). Blended Learning. *Journal of Turkish Science Education, 8*(2), 3-8.

Albanese, M. A., & Mitchell, S. (1993). Problem-based learning: A review of literature on its outcomes and implementation issues. *Academic Medicine, 68*(1), 52–81. doi:10.1097/00001888-199301000-00012 PMID:8447896

Albion, P. R. (2008). Web 2.0 in teacher education: Two imperatives for action. *Computers in Schools, 25*(3-4), 181–198. doi:10.1080/07380560802368173

Alexander, B. (2004). Going Nomadic: Moblile Learning in Higher Education. *EDUCAUSE Review, 39*(5).

Allred, C. R., Fawcett, S. E., Wallin, C., & Magnan, G. M. (2011). A Dynamic Cooperation Capability as a Source of Competitive Advantage. *Decision Sciences, 42*(1), 129–161. doi:10.1111/j.1540-5915.2010.00304.x

Altınkurt, Y., & Yılmaz, K. (2010). Değerlere Göre Yönetim ve Örgütsel Adalet İlişkisinin Ortaöğretim Okulu Öğretmenlerinin Algılarına Göre İncelenmesi. *Kuram ve Uygulamada Eğitim Yönetimi., 16*(4), 463–484.

Amini, M., Blair, M. A., Forrester, J., Goldstein, S. J., Katsouros, M., Rocchio, R. A., & Williams, A., III. (2013). *Developing a campus mobile strategy: Guidelines, tools, and best practices.* Retrieved from https://library.educause.edu/-/media/files/library/2013/1/acti1303-pdf.pdf

Andersen, P. (2007). *What is Web 2.0?: ideas, technologies and implications for education.* Bristol: JISC.

Annand, D. (2007). Re-organizing universities for the information age. *International Review of Research in Open and Distance Learning, 8*(3), 1–9. doi:10.19173/irrodl.v8i3.372

Anonymous. (2002). Managing by values. *Strategic Dimension*, (22).

APICS. (2019). *Supply Chain Operations Reference (SCOR) model.* https://www.apics.org/apics-for-business/frameworks/scor

Apple. (2010). *Challenge-Based Learning. A classroom guide.* Retrieved from https://images.apple.com/education/docs/CBL_Classroom_Guide_Jan_2011.pdf

Arai, S., Sakamoto, K., & Washizaki, H. (2014). *A Gamified Tool for Motivating Developers to Remove Warnings of Bug Pattern Tools.* Paper presented at the IWESEP 2014, Osaka, Japan. 10.1109/IWESEP.2014.17

Arguedas-Matarrita, C., Concari, S. B., García-Zubia, J., & Marchisio, S. T. (2017). *A teacher training workshop to promote the use of the VISIR Remote Laboratory for electrical circuits teaching.* Presented at the 4th Experiment@ International Conference (exp.at'17), Portugal. 10.1109/EXPAT.2017.7984351

Arnab, S., Berta, R., Earp, J., de Freitas, S., Popescu, M., Romero, M., . . . Usart, M. (2012). Framing the Adoption of Serious Games in Formal Education. *Electronic Journal of e-Learning, 10*(2), 59-171. Available online at www.ejel.com

Asif, M., & Krogstie, J. (2011). Mobile student information system. *Campus-Wide Information Systems, 28*(1), 5–15. doi:10.1108/10650741111097269

Asociación Española de Normalización, U. N. E. (2018). *Especificación UNE 0060:2018.* Retrieved from https://www.une.org/encuentra-tu-norma/busca-tu-norma/norma?c=N0060640

Astin, A. (1993). *What Matters in College: Four Critical Years Revisited.* San Francisco: Jossey-Bass.

Autor, D., Levy, F., & Murnane, R. (2003). The skill content of recent technological change: An empirical exploration. *The Quarterly Journal of Economics, 118*(4), 1279–1333. doi:10.1162/003355303322552801

Bagheri, M., & Haghighi, M. (2016). The Effect of the Internet of Things (IoT) on Education Business Model. In *2016 12th International Conference on Signal-Image Technology & Internet-Based Systems (SITIS).* IEEE Computer Society.

Baker, R., & Associates. (2010). Data mining for education. International Encyclopedia of Education, 7(3), 112–118.

Barahate, S. R. (2012). *Educational data mining as a trend of data mining in educational system.* International Conference & Workshop on Recent Trends in Technology (TCET), Mumbai, India.

Barr, M. (2018). Student attitudes to games-based skills development: Learning from video games in higher education. *Computers in Human Behavior, 80*, 283–294. doi:10.1016/j.chb.2017.11.030

Becker, A., Cummins, M., Davis, A., Freeman, A., Hall Giesinger, C., & Ananthanarayanan, V. (2017). *NMC Horizon Report: 2017 Higher Education Edition.* Austin, TX: The New Media Consortium.

Beck, V. (2014). The effects of the implementation of value-based management. *International Journal of Economic Sciences and Applied Research., 7*(2), 153–165.

Bell, J., & Harrison, B. T. (2018). *Vision and Values in Managing Education: Successful Leadership Principles and Practice.* Abingdon, UK: Published by Routledge. doi:10.4324/9781351041508

Bellotti, F., Berta, R., De Gloria, A., Lavagnino, E., Antonaci, A., Dagnino, F., ... Mayer, I. S. (2014). Serious games and the development of an entrepreneurial mindset in higher education engineering students. *Entertainment Computing, 5*(4), 357–366. doi:10.1016/j.entcom.2014.07.003

Belski, I., Adunka, R., & Mayer, O. (2016). Educating a creative engineer: Learning from engineering professionals. *Procedia CIRP, 39*, 79–84. doi:10.1016/j.procir.2016.01.169

Benedek, A., & Molnar, G. (2015, July). *E-teaching and Digitalization at BME.* Paper presented at the 19th International Conference on Engineering Education, Zagreb.

Berhad, M. (2015). *National Internet of Things (IoT) Strategic Roadmap.* Available: http://www.mimos.my/IoT/National_IoT_Strategic_Roadmap_Summary.pdf

Blanchard, K. (1997). *Managing by values.* Executive Excellence.

Botta, A., & de Donato, W. (2014). On the Integration of Cloud Computing and Internet of Things. In *Proceedings of the International Conference on Future Internet of Things and Cloud* (pp. 23 – 30). IEEE Conference Publications. 10.1109/FiCloud.2014.14

Bower, M. (2016). Deriving a typology of Web 2.0 learning technologies. *British Journal of Educational Technology*, *47*(4), 763–777. doi:10.1111/bjet.12344

Bower, M., Hedberg, J. G., & Kuswara, A. (2010). A framework for Web 2.0 learning design. *Educational Media International*, *47*(3), 177–198. doi:10.1080/09523987.2010.518811

Braghirolli, L., Ribeiro, J., Weise, A., & Pizzolato, M. (2016). Benefits of educational games as an introductory activity in industrial engineering education. *Computers in Human Behavior*, *58*, 315–324. doi:10.1016/j.chb.2015.12.063

Branch, R. M. (2009). Instructional design: The ADDIE approach. Springer Science & Business Media.

Branco, M. V., & Coelho, L. A. (2017). *Aspectos de diferenciação entre laboratórios remotos e simuladores.* Presented at the Congresso Brasileiro de Educação em Engenharia - COBENGE, Brazil.

Brazil. (2017). *Lei n. 14.363 de 25 de janeiro de 2008.* Retrieved from http://www.leisestaduais.com.br/sc/lei-ordinaria-n-14363-2008-santa-catarina-dispoe-sobre-a-proibicao-do-uso-de-telefone-celular-nas-escolas-estaduais-do-estado-de-santa-catarina

Breuer, J., & Bente, G. (2010). Why So Serious? On the Relation of Serious Games and Learning. *Eludamos (Göttingen)*, *4*(1), 7–24.

Briggs, B., Foutty, J., & Hodgetts, C. (2016). *Tech Trends.* Deloitte University Press.

Bukvić, V. (2016). Value-based management with some practical examples in Slovenian industries. *Advanced in Business-Related Scientific Research Journal, 7*(2).

Buniyamin, N., Mat, U. B., & Arshad, P. M. (2016). Educational data mining for prediction and classification of engineering students achievement. In *IEEE 7th International Conference on Engineering Education*. IEEE.

Butler, J. W. (2012). Grappling with change: Web 2.0 and teacher educators. In D. Polly, C. Mims, & K. Persichitte (Eds.), *Developing Technology-Rich Teacher Education Programs: Key Issues* (pp. 135–150). Hershey, PA: IGI Global. doi:10.4018/978-1-4666-0014-0.ch010

Calders, T., & Pechenizkiy, M. (2012). Introduction to the Special Section on Educational Data Mining. *SIGKDD Explorations*, *13*(2), 3–6. doi:10.1145/2207243.2207245

Carini, R. M., Kuh, G. D., & Klein, S. P. (2006). Student engagement and student learning: Testing the linkages. *Research in Higher Education*, *47*(1), 1–32. doi:10.100711162-005-8150-9

Carrara, A. C., Silva, I. N., Lotthammer, K. S., Silva, J. B., & Ferenhof, H. (2018). *O uso do laboratório remoto VISIR em oficinas de eletrônica para estudantes do ensino médio*. Presented at the Congresso Brasileiro de Educação em Engenharia - COBENGE, Brazil.

Castells, M. (2004). Informationalism, networks, and the network society: a theoretical blueprint. In M. Castells (Ed.), *Network Society: A cross-cultural perspective* (pp. 3–45). Cheltenham, UK: Edward Elgar Publishing Limited. doi:10.4337/9781845421663.00010

Catalano, G. D., & Catalano, K. (1999). Transformation: From teacher-centered to student-centered engineering education. *Journal of Engineering Education, 88*(1), 59–64. doi:10.1002/j.2168-9830.1999.tb00412.x

Chan, J., Chen, Z., Cormane, I., Her, N., & Thomas, R. (2006). *Cell Phone Industry Analysis Report.* Sacremanto State University Press.

Chassignol, M., Khoroshavin, A., & Bilyatdinova, A. (2018). Artificial Intelligence trends in education: A narrative overview. *Procedia Computer Science, 136*, 16–24. doi:10.1016/j.procs.2018.08.233

Chen, H. L., Lattuca, L. R., & Hamilton, E. R. (2008). Conceptualizing engagement: Contributions of faculty to student engagement in engineering. *Journal of Engineering Education, 97*(3), 339–353. doi:10.1002/j.2168-9830.2008.tb00983.x

Chen, J., Wang, M., Kirschner, P. A., & Tsai, C. C. (2018). The role of collaboration, computer use, learning environments, and supporting strategies in CSCL: A meta-analysis. *Review of Educational Research, 88*(6), 799–843. doi:10.3102/0034654318791584

Cheung, L. (2016). Using the ADDIE model of instructional design to teach chest radiograph interpretation. *Journal of Biomedical Education, 2016*, 1–6. doi:10.1155/2016/9502572

Chickering, A. W., & Gamson, Z. F. (1987). Seven principles for good practice in undergraduate education. *AAHE Bulletin, 3*, 7.

Chis, A. E., Moldovan, A. N., Murphy, L., Pathak, P., & Muntean, C. H. (2018). Investigating flipped classroom and problem-based learning in a Programming Module for Computing Conversion course. *Journal of Educational Technology & Society, 21*(4), 232–247.

Chun, S. G., Chung, D., & Shin, Y. (2013). Are Students Satisfied with The Use of Smartphone Apps. *Issues in Information Systems, 14*(2), 23–33.

Cisco. (2013). *Industrial smart solutions: connecting the factory to the enterprise.* Whitepaper. Available: https://www.cisco.com/c/dam/global/en_ph/assets/tomorrow-starts-here/files/white_paper.pdf

Coller, B., & Scott, M. (2009). Effectiveness of using a video game to teach a course in mechanical engineering. *Computers & Education, 53*(3), 900–912. doi:10.1016/j.compedu.2009.05.012

Corter, J. E., Esche, S. K., Chassapis, C., Ma, J., & Nickerson, J. V. (2011). Process and learning outcomes from remotely-operated, simulated, and hands-on student laboratories. *Computers & Education, 57*(3), 2055–2067. doi:10.1016/j.compedu.2011.04.009

Cotter, J., Forster, G., & Sweeney, E. (2009). Supply Chain Learning : The Role of Games. *Journal of the National Institute for Transport and Logistics, 10*(3), 32–36.

Creswell, J. W. (2013). *Research Design (International Student Edition): Qualitative, Quantitative, and Mixed Methods Approaches* (4th ed.). Sage Publications, INC.

Crookall, D. (2010). Serious Games, Debriefing, and Simulation/Gaming as a Discipline. *Simulation & Gaming, 41*(6), 898–920. doi:10.1177/1046878110390784

Damiani, L., Demartini, M., Guizzi, G., Revetria, R., & Tonelli, F. (2018). Augmented and virtual reality applications in industrial systems: A qualitative review towards the industry 4.0 era. *IFAC-PapersOnLine, 51*(11), 624–630. doi:10.1016/j.ifacol.2018.08.388

Daniels, J., Sargolzaei, S., Sargolzaei, A., Ahram, T., Laplante, P. A., & Amaba, B. (2018). The internet of things, artificial intelligence, blockchain, and professionalism. *IT Professional, 20*(6), 15-19. doi:10.1109/MITP.2018.2875770

Dávidekova, M., Mjartan, M., & Greguša, A. (2017). Utilization of Virtual Reality in Education of Employees in Slovakia. *The 8th International Conference on Emerging Ubiquitous Systems and Pervasive Networks*. Available: https://pdf.sciencedirectassets.com

De Dreu, C. K. W., & Beersma, B. (2010). Fast track reports: Team confidence, motivated information processing, and dynamic group decision making. *European Journal of Social Psychology, 40*(7), 1110–1119. doi:10.1002/ejsp.763

De Vries, P., Klaasse, R., & Kamp, A. (2017). Emerging technologies in engineering education: can we make it work? *Proceedings of 13th International CDIO Conference*, 1-12.

Deif, A. (2017). Insights on lean gamification for higher education. *International Journal of Lean Six Sigma, 8*(3), 359–376. doi:10.1108/IJLSS-04-2016-0017

Demirtaş, Z. & Kılıç, M.F. (2015). Okul yöneticilerinin değerlerle yönetim kavramına ilişkin algıları. *Turkish Journal of Educational Studies*, (2).

Deterding, S., Dixon, D., Khaled, R., & Nacke, L. (2011). *From Game Design Elements to Gamefulness: Defining "Gamification"*. Paper presented at the 15th International Academic MindTrek Conference, Tampere. 10.1145/2181037.2181040

Dias, J. (2017). Teaching operations research to undergraduate management students: The role of gamification. *International Journal of Management Education, 15*(1), 98–111. doi:10.1016/j.ijme.2017.01.002

Dijk, T. V., Spil, T., Burg, S. V., Wenzler, I., & Dalmolen, S. (2015). Present or Play, Some Evidence on the Effect on Behaviour of Serious Gaming. *International Journal of Game-Based Learning, 5*(2), 55–69. doi:10.4018/ijgbl.2015040104

Dobrescu, L., Greiner, B., & Motta, A. (2015). Learning economics concepts through game-play: An experiment. *International Journal of Educational Research, 69*, 23–37. doi:10.1016/j.ijer.2014.08.005

Doğan, S. (2016). Model of values-based management process in school: A mixed design study. *Sciedu Press., 5*(1), 83–96.

Doherty, J. J., Hansen, M. A., & Kaya, K. K. (2005). Teaching information skills in the information age: The need for critical thinking. *Library Philosophy and Practice*, *1*(2), 1–9.

Dokuwiki. (2019). *About Dokuwiki*. Retrieved from https://www.dokuwiki.org/pt-br:dokuwiki

Dolan, S. L., & Richley, B. A. (2006). MBV: a new order philosophy for a new economic order. Handbook of Business Strategy.

Dolan, S. L., & Altman, Y. (2012). Managing by values: The leadership spirituality connection. *People and Strategy*, *35*(4), 20–26.

Dolan, S. L., & Garcia, S. (1999). *La Gestion par Valeurs: Una Nowelle Culture pour les Organisations*. Montreal: Editions Nouvelles.

Dolan, S. L., & Garcia, S. (2002). Managing by values cultural redesign for strategic organizational change at the dawn of the twenty-first century. *Journal of Management Development*, *21*(2), 101–117. doi:10.1108/02621710210417411

Dolan, S. L., & Raich, M. (2013). Coaching by values, entrepreneurship, and care A framework for reengineering an innovative and sustainable culture. *Kindai Management Review*, *1*, 2186–6961.

Dolmans, D. H., Loyens, S. M., Marcq, H., & Gijbels, D. (2016). Deep and surface learning in problem-based learning: A review of the literature. *Advances in Health Sciences Education: Theory and Practice*, *21*(5), 1087–1112. doi:10.100710459-015-9645-6 PMID:26563722

Dosheela, D., & Binod, K. (2019). *Exploring the Internet of Things (IoT) in Education: A Review*. Information Systems Design and Intelligent Applications. Available https://link.springer.com/chapter/10.1007%2F978-981-13-3338-5_23

Downes-Le Guin, T., Baker, R., Mechling, J., & Ruyle, E. (2012). Myths and realities of Driskell, J., & Dwyer, D. (1984). Microcomputer Videogame Based Training. *Educational Technology*, *24*(2), 11–16. Retrieved from http://www.jstor.org/stable/44427307

Duffy, T. C., & Cunningham, D. D. (1996). *Constructivism: Implications for the design and delivery of instruction. In Handbook of Research for Educational Communications and Technology*. New York: MacMillan Library.

Durak, G., & Ataizi, M. (2016). The ABC's of Online Course Design According to Addie Model. *Universal Journal of Educational Research*, *4*(9), 2084–2091. doi:10.13189/ujer.2016.040920

Dutra Moresi, E. A., Oliveira Braga Filho, M., Alves Barbosa, J., Carmo Lopes, M., Alves Tito de Morais, M. A., Alves dos Santos, J. C., . . . Osmala Júnior, W. A. (2017). O emprego do aprendizado baseado em desafios no desenvolvimento de aplicativos móveis [The use of Challenge Based Learning in mobile application development]. *2017 12th Iberian Conference on Information Systems and Technologies (CISTI)*, 1-6.

Einsenberg, M. B. (2008). Information literacy: Essential skills for the information age. *DESIDOC Journal of Library and Information Technology*, *28*(2), 39–47. doi:10.14429/djlit.28.2.166

Elvekrok, I., Veflen, N., Nilso, E., & Gausdal, A. (2017). Firm innovation benefits from regional triple-helix networks. *Regional Studies, 52*(9), 1214–1224. doi:10.1080/00343404.2017.1370086

Englund, C., Olofsson, A. D., & Price, L. (2017). Teaching with technology in higher education: Understanding conceptual change and development in practice. *Higher Education Research & Development, 36*(1), 73–87. doi:10.1080/07294360.2016.1171300

Entertainment Software Association. (2015). *2015: Sales, Demographic and Usage Data - essential Facts about the Computer and Video Game Industry*. Retrieved from http://www.theesa.com/wp-content/uploads/2015/04/ESA-Essential-Facts-2015.pdf

Etzkowitz, H., & Leydesdorff, L. (1995). The Triple Helix -- University-Industry-Government Relations: A Laboratory for Knowledge Based Economic Development. *EASST Review, 14*(1), 14-19. Retrieved from https://ssrn.com/abstract=2480085

Etzkowitz, H. (2003). Innovation in Innovation: The Triple Helix of University-Industry-Government Relations. *Social Sciences Information. Information Sur les Sciences Sociales, 43*(3), 293–337. doi:10.1177/05390184030423002

European Commission. (2017). *Germany: Industrie 4.0. Digital transformation monitor*. Retrieved from https://ec.europa.eu/growth/tools-databases/dem/monitor/sites/default/files/DTM_Industrie%204.0.pdf

European Commission. (2018). *The Factories of the Future*. https://ec.europa.eu/digital-single-market/en/factories-future

Everson, K. (2015). Learning Is All in the Wrist. *Chief Learning Officer, 14*(4), 18–21.

Faieza, A., Azreen, A., & Rohidatun, M. (2015). Virtual Reality Application In Design, Prototype and Test Drive Processes In Automotive Industry. *International Journal of Advance Research In Science and Engineering, 4*(1). Available: https://pdfs.semanticscholar.org/d485/2354c2349bfe2a0aad2919587a8675071dee.pdf

Farina, J., Marchisio, S. T., Concari, S. B., Lerro, F. G., Pozzo, M. I., Alves, G. R. C., … Nilsson, K. (2017). *Preparando estudiantes secundarios para carreras de ingenieria: un estudio de caso utilizando el laboratorio remoto VISIR*. Presented at the Congresso Brasileiro de Educação em Engenharia - COBENGE.

Farzan, R., & Brusilovsky, P. (2011). Encouraging user participation in a course recommender system: An impact on user behavior. *Computers in Human Behavior, 27*(1), 276–284. doi:10.1016/j.chb.2010.08.005

Felder, R. M. (2012). Engineering education: A tale of two paradigms. In B. McCabe, M. Pantazidou, & D. Phillips (Eds.), *Shaking the Foundations of Geo-Engineering Education* (pp. 9–14). Boca Raton, FL: CRC Press. doi:10.1201/b15096-4

Feng, Y., D' Amours, S., & Beauregard, R. (2010). Simulation and performance evaluation of partially and fully integrated sales and operations planning. *International Journal of Production Research, 48*(19), 5859–5883. doi:10.1080/00207540903232789

Fernandes, J., Duarte, D., Ribeiro, C., Farinha, C., Pereira, J. M., & Silva, M. M. (2012). iThink: A game-based approach towards improving collaboration and participation in requirement elicitation. *Procedia Computer Science, 15*, 66-77. doi:10.1016/j.procs.2012.10.059

Fidalgo, A. V., Alves, G. R., Alves, M. A., & Viegas, M. C. (2014). *Adapting Remote Labs to Learning Scenarios: Case Studies Using VISIR and RemotElectLab*. IEEE Revista Iberoamericana de Tecnologias Del Aprendizaje.

Fiore, S. M., Graesser, A., Greiff, S., Griffin, P., Gong, B., Kyllonen, P., … von Davier, A. (2017). *Collaborative problem solving: Considerations for the national assessment of educational progress*. Retrieved from https://nces.ed.gov/nationsreportcard/pdf/researchcenter/collaborative_problem_solving.pdf

Förster, A., Dede, J., Könsgen, A., Udugama, A., & Zaman, I. (2017). Teaching the Internet of Things. GetMobile. *Mobile Computing and Communications, 20*(3), 24–28.

Franklin, T., & Van Harmelen, M. (2007). *Web 2.0 for content for learning and teaching in higher education.* Retrieved from https://staff.blog.ui.ac.id/harrybs/files/2008/10/web-2-for-content-for-learning-and-teaching-in-higher-education.pdf

Fruchter, R. (2001). Dimensions of teamwork education. *International Journal of Engineering Education, 17*(4-5), 426–430.

Garris, R., Ahlers, R., & Driskell, J. E. (2002). Games, motivation, and learning: A research and practice model. *Simulation & Gaming, 33*(4), 441–467. doi:10.1177/1046878102238607

Gee, J. P. (2007). *Good video games and good learning: Collected essays on video games, learning, and literacy*. New York: Peter Lang International Academic Publishers. doi:10.3726/978-1-4539-1162-4

Greenhow, C., Robelia, B., & Hughes, J. E. (2009). Learning, teaching, and scholarship in a digital age: Web 2.0 and classroom research: What path should we take now? *Educational Researcher, 38*(4), 246–259. doi:10.3102/0013189X09336671

Green, M., Cantu, A., & Wardle, A. (2014). A study examining student preferences of mobile applications at an institution of higher learning in South Texas. In T. Bastiaens (Ed.), *Proceedings of World Conference on E-Learning* (pp. 717-723). New Orleans, LA: Association for the Advancement of Computing in Education (AACE).

Griffin, P., Care, E., & McGaw, B. (2012). The changing role of education and schools. In P. Griffin, E. Care, & B. McGaw (Eds.), *Assessment and teaching of 21st century skills* (pp. 1–15). New York: Springer. doi:10.1007/978-94-007-2324-5_1

Gros, B., & López, M. (2016). Students as co-creators of technology-rich learning activities in higher education. *International Journal of Educational Technology in Higher Education, 13*(1), 28. doi:10.118641239-016-0026-x

Grosseck, G. (2009). To use or not to use web 2.0 in higher education? *Procedia: Social and Behavioral Sciences, 1*(1), 478–482. doi:10.1016/j.sbspro.2009.01.087

Gubbi, J., Buyya, R., Marusic, S., & Palaniswamia, M. (2012). *Internet of Things (IoT): A Vision, Architectural Elements, and Future Directions*. Available: https://arxiv.org/ftp/arxiv/papers/1207/1207.0203.pdf

Guillén-Nieto, V., & Aleson-Carbonell, M. (2012). Serious games and learning effectiveness: The case of It's a Deal! *Computers & Education*, *58*(1), 435–448. doi:10.1016/j.compedu.2011.07.015

Güldal, H., & Çakıcı, Y. (2017). Educational Data Mining, Balkan. *Educational Studies*, *1*, 135–143.

Gupta, P., Gop, K., & Kyei-Blankson, L. (2014). College students' usage of and expectations from university owned mobile applications. In T. Bastiaens (Ed.), *Proceedings of World Conference on E-Learning* (pp. 742-745). New Orleans, LA: Association for the Advancement of Computing in Education (AACE).

Gustafsson, A., Katzeff, C., & Bang, M. (2009). Evaluation of a pervasive game for domestic energy engagement among teenagers. Computers in Entertainment. *CIE*, *7*(4), 54. doi:10.1145/1658866.1658873

Gustavsson, I., Nilsson, K., Zackrisson, J., Garcia-Zubia, J., Hernandez-Jayo, U., Nafalski, A., ... Hkansson, L. (2009). N objectives of instructional laboratories, individual assessment, and use of collaborative remote laboratories. *IEEE Transactions on Learning Technologies*, *2*(4), 263–274. doi:10.1109/TLT.2009.42

Gustavsson, I., Zackrisson, J., & Lundberg, J. (2014). *VISIR work in progress*. Presented at the *IEEE Global Engineering Education Conference (EDUCON)*, Turkey. 10.1109/EDUCON.2014.6826253

Hair, J. F., Anderson, R. E., Tatham, R. L., & Black, W. C. (1995). *Multivariate Data Analysis with Readings* (4th ed.). Prentice-Hall Internatimal, Inc.

Hamari, J. (2013). Transforming homo economicus into homo ludens: A field experiment on gamification in a utilitarian peer-to-peer trading service. Electronic Commerce Research and Applications. *Social Commerce*, *12*(Part 2), 236–245. doi:10.1016/j.elerap.2013.01.004

Hamari, J. (2015). Do badges increase user activity? A field experiment on the effects of gamification. *Computers in Human Behavior*. doi:10.1016/j.chb.2015.03.036

Harung, H. S., & Dahl, T. (1995). Increased productivity and quality through management by values: A case study of Manpower Scandinavia. *The TQM Magazine*, *7*(2), 13–22. doi:10.1108/09544789510081063

Hauge, J. M. B., Pourabdollahian, B., & Riedel, J. C. K. H. (2013). The Use of Serious Games in the Education of Engineers. In C. Emmanouilidis, M. Taisch, & D. Kiritsis (Eds.), *Advances in Production Management Systems. Competitive Manufacturing for Innovative Products and Services. APMS 2012. IFIP Advances in Information and Communication Technology* (Vol. 397). Berlin: Springer. doi:10.1007/978-3-642-40352-1_78

Haynes, F. (1998). *Ethical School*. London: Routledge Press.

Hense, J., & Mandl, H. (2014). Learning in or with games? Quality criteria for digital learning games from the perspectives of learning, emotion, and motivation theory. In D. G. Sampson, D. Ifenthaler, J. M. Spector, & P. Isaias (Eds.), Digital systems for open access to formal and informal learning (pp. 181-193). Piräus: Springer. doi:10.1007/978-3-319-02264-2_12

Hill, G. W. (1982). Group Versus Individual Performance: Are N + 1 Heads Better Than One? *Psychological Bulletin, 91*(3), 517–539. doi:10.1037/0033-2909.91.3.517

Hoffmann, L. (2009). Learning through games. *Communications of the ACM, 52*(8), 21–22. doi:10.1145/1536616.1536624

Hong, J.-C., Cheng, C.-L., Hwang, M.-Y., Lee, C.-K., & Chang, H.-Y. (2009). Assessing the educational values of digital games. *Journal of Computer Assisted Learning, 25*(5), 423–437. doi:.00319.x doi:10.1111/j.1365-2729.2009

Huang, W. D., Johnson, T. E., & Han, S.-H. C. (2013). Impact of online instructional game features on college students' perceived motivational support and cognitive investment: A structural equation modeling study. *The Internet and Higher Education, 17*, 58–68. doi:10.1016/j.iheduc.2012.11.004

Huba, M., & Kozák, Š. F. (2016). From E-learning to Industry 4.0. *International Conference on Emerging eLearning Technologies and Applications (ICETA)*, 3-108. DOI: 10.1109/ICETA.2016.7802083

Hummel, H. G. K., van Houcke, J., Nadolski, R. J., van der Hiele, T., Kurvers, H., & Löhr, A. (2011). Scripted cooperation in serious gaming for complex learning: Effects of multiple perspectives when acquiring water management skills. *British Journal of Educational Technology, 42*(6), 1029–1041. doi:.01122.x doi:10.1111/j.1467-8535.2010

Huy, N. P., & Van Thanh, D. (2012). Evaluation of Mobile App Paradigms. *10th International Conference on Advances in Mobile Computing & Multimedia*, 25-30.

Hwang, G.-J., Sung, H.-Y., Hung, C.-M., Huang, I., & Tsai, C.-C. (2012). Development of a personalized educational computer game based on students' learning styles. *Educational Technology Research and Development, 60*(4), 623–638. doi:10.100711423-012-9241-x

IIBA, (2015). *A Guide to the Business Analysis Body of Knowledge (BABOK Guide)*. International Institute of Business Analysis (IIBA).

Ipsos. (2011). *Engagement Unleashed: Gamification for Business, Brands and Loyalty*. http://saatchi.com/en-us/news/engagement_unleashed_gamification_for_business_brands_and_loyalty

Isoc, D., & Surubaru, T. (2018). *Engineering Education Using Professional Activity Simulators*. Available https://link-springer-com.ezproxy.utm.my/chapter/10.1007/978-3-030-11932-4_50

ITU Metallurgical and Materials Engineering Undergraduate Program. (2019). *Program educational objectives*. Retrieved from http://mme.itu.edu.tr/en/abet/educational-objectives

ITU Metallurgical and Materials Engineering Undergraduate Program. (2019). *Student Outcomes*. Retrieved from http://mme.itu.edu.tr/en/abet/student-outcomes

ITU School of Foreign Languages. (2019). *Minimum English scores for undergraduate programs.* Retrieved from http://www.ydy.itu.edu.tr/en/to-be-exempt-from-program/

ITU. (2012). *Overview of the Internet of Things.* Retrieved from: http://www.itu.int/ITUT/recommendations/rec.aspx?rec=Y.2060

ITU. (2019). *ITU history.* Retrieved from http://global.itu.edu.tr/global-itu/history

ITU. (2019). *ITU vision and mission.* Retrieved from http://global.itu.edu.tr/global-itu/vision-and-mission

Jaakson, K. (2010). Management by values: Are some values important than others? *Journal of Management Development, 29*(9), 795–806. doi:10.1108/02621711011072504

Janssen, D., Tummel, C., Richert, A., & Isenhardt, I. (2016). Higher Education – Immersion as a Key Construct for Learning 4.0. *International Journal of Advanced Corporate Learning, 9*(2), 20–26. doi:10.3991/ijac.v9i2.6000

Jaschik, S., & Lederman, D. (2015). *The 2016 inside Higher Ed Survey of College and University Chief Academic Officers.* Washington, DC: Inside Higher Ed and Gallup. Available: https://www.insidehighered.com/booklet/survey-college-and-university-chief-academic-officers

Johnson, D., Means, T., & Khey, D. N. (2013). *A state of flux: Results of a mobile device surveyat the University of Florida.* EDUCAUSE.

Johnson, L. F., Smith, R. S., Smythe, J. T., & Varon, R. K. (2009). *Challenge-Based Learning: An approach for our time.* Austin, TX: The New Media Consortium.

Jonassen, D. H. (1999). Designing constructivist learning environments. In C. M. Reigeluth (Ed.), *Instructional-Design Theories and Models* (Vol. 2, pp. 215–239). Lawrence Erlbaum Associates.

Jones, B. A., Madden, G. J., & Wengreen, H. J. (2014). The FIT game: Preliminary evaluation of a gamification approach to increasing fruit and vegetable consumption in school. *Preventive Medicine, 68*, 76-79. doi:10.1016/j.ypmed.2014.04.015

Kamp, A. (2016). Engineering Education in a Rapidly Changing World; Rethinking the Vision for Higher Education (2nd ed.). 4TU-Centre for Engineering Education.

Kang, Y., Han, M., Han, K., & Kim, J. (2015). A Study on the Internet of Things (IoT) Applications. *International Journal of Software Engineering and Its Applications, 9*(9), 117–126. doi:10.14257/ijseia.2015.9.9.10

Kano, N., Seraku, N., Takahashi, F., & Tsuji, S. (1984). Attractive quality and Must-be quality. *The Journal of the Japanese Society for Quality Control, 14*(2), 39–48.

Kapp, K. M. (2012). *The gamification of learning and instruction: Game-based methods and strategies for training and education.* San Francisco: Pfeiffer.

Kastenberg, W. E., Hauser-Kastenberg, G., & Norris, D. (2006). *Proceedings of 36th ASEE/IEEE Frontiers in Education Conference.* IEEE.

Kasurinen, J., & Knutas, A. (2018). Publication trends in gamification: A systematic mapping study. *Computer Science Review*, *27*, 33–44. doi:10.1016/j.cosrev.2017.10.003

Katharina M., Dominic G. (2015). In-Factory Learning-Qualification for the Factory of the Future. *Acta Universitatis Cibiniensis – Technical Series 2015, 66*, 159-164. Doi:10.1515/aucts-2015-0046

Kaukalias, T., & Chatzimisios, P. (2015). *Internet of Things (IoT)*. IGI Global. Available: https://www.irma-international.org/viewtitle/112465/

Kerwin, S. & MacLein, J. & Bell-Laroche, D. (2014). The value of managing by values a Canadian sports organization case study. *Journal of Applied Sport Management*, *16*(4).

Kim, B. (2012). Harnessing the power of game dynamics1: Why, how to, and how not to gamify the library experience. *College & Research Libraries News*, *73*(8), 465–469. doi:10.5860/crln.73.8.8811

Kin, S., & Kam, C. (2015). The Impact of the Internet of Things on Engineering Education. Academic Press.

King, J. & South, J. (2017). *Reimagining the Role of Technology in Higher Education: A Supplement to the National Education Technology Plan*. U.S. Department of Education, Office of Educational Technology.

Klein, D. (2012, Feb.). How to decide: Mobile websites vs. mobile apps. *Inspire*.

Klopfer, J. (2016). *Augmented Learning: Research and Design of Mobile Education Games*. MIT Press.

Kolb, D. A. (2014). *Experiential learning: Experience as the source of learning and development*. FT Press.

Konsky, B. R., Miller, C., & Jones, A. (2016). The skills framework for the information age: Engaging stakeholders in curriculum design. *Journal of Information Systems Education*, *27*(1), 37–49.

Korucu, A. T., & Cakir, H. (2018). The effect of dynamic web technologies on student academic achievement in problem-based collaborative learning environment. *Malaysian Online Journal of Educational Technology*, *6*(1), 92–108.

Kotsiantis, S. (2009). Educational Data Mining: A Case Studyfor Predicting Dropout- Prone Students. *International Journal of Knowledge Engineering and Soft Data Paradigms*, *1*(2), 101–111. doi:10.1504/IJKESDP.2009.022718

Krzakıewıcz, K. (2012). Management by values – a strategic dimension. Management, 16(2).

Krzakıewıcz, K. (2012). Management by values – a strategic dimension. *Management*, *16*(2), 7–15. doi:10.2478/v10286-012-0051-3

Ku, H.-Y., Tseng, H. W., & Akarasriworn, C. (2013). Cooperation factors, teamwork satisfaction, and student attitudes toward online collaborative learning. *Computers in Human Behavior*, *29*(3), 922–929. doi:10.1016/j.chb.2012.12.019

Kuh, G. D., Cruce, T. M., Shoup, R., Kinzie, J., & Gonyea, R. M. (2008). Unmasking the effects of student engagement on first-year college grades and persistence. *The Journal of Higher Education*, *79*(5), 540–563. doi:10.1080/00221546.2008.11772116

Kumar, J. (2013). Gamification at Work: Designing Engaging Business Software. Design, User Experience, and Usability. Health, Learning, Playing, Cultural, and Cross-Cultural User Experience. *Lecture Notes in Computer Science*, *8013*, 528–537. doi:10.1007/978-3-642-39241-2_58

Laird, T. F. N., & Kuh, G. D. (2005). Student experiences with information technology and their relationship to other aspects of student engagement. *Research in Higher Education*, *46*(2), 211–233. doi:10.100711162-004-1600-y

Lambert, D., & Burduroglu, R. (2000). Measuring and Selling the Value of Logistics. *International Journal of Logistics Management*, *11*(1), 1–18. doi:10.1108/09574090010806038

Landers, R. N., & Landers, A. K. (2014). An Empirical Test of the Theory of Gamified Learning: The Effect of Leaderboards on Time-on-Task and Academic Performance. *Simulation & Gaming*, *45*(6), 769-785. doi:10.1177/1046878114563662

Lavechia, J., Silva, J. B., & Spanhol, F. J. (2017). *Publicações científicas do Laboratório de Experimentação Remota – RExLab: uma revisão sistemática*. Informação & Informação.

Lee, J. J., & Hammer, J. (2011). Gamification in education: What, how, why bother? *Academic Exchange Quarterly*, *15*(2), 146.

Lee, J., Luchini, K., Michael, B., Norris, C., & Soloway, E. (2004, April). More than just fun and games: Assessing the value of educational video games in the classroom. In *CHI'04 extended abstracts on Human factors in computing systems* (pp. 1375–1378). ACM. doi:10.1145/985921.986068

Lenschow, R. J. (1998). From teaching to learning: A paradigm shift in engineering education and lifelong learning. *European Journal of Engineering Education*, *23*(2), 155–161. doi:10.1080/03043799808923494

Lensing, K., & Friedhoff, J. (2018). Designing a curriculum for the Internet-of-Thing- Laboratory to foster creativity and marker mindset within varying target group. *8th conference on learning factories 2018- Advance Engineering Education and Training for Manufacturing Innovation*. www.sciencedirect.com

Lima, N., Viegas, C., Alves, G., & Garcia-Peñalvo, F. J. (2016). *VISIR usage as an educational resource*. Presented at the TEEM '16, Spain.

Liu, Y., Alexandrova, T., & Nakajima, T. (2011). Gamifying intelligent environments. In *Proceedings of the 2011 international ACM workshop on ubiquitous meta user interfaces* (pp. 7-12). ACM. 10.1145/2072652.2072655

Livingstone, K. A. (2015). *The impact of Web 2.0 in Education and its potential for language learning and teaching*. Retrieved from https://www.researchgate.net/publication/272487634_The_impact_of_Web_20_in_Education_and_its_potential_for_language_learning_and_teaching

Lobo, M. C. C. (2011). *Using Remote Experimentation in a Large Undergraduate Course: Initial Findings*. Presented at the *41st Asee/ieee Frontiers In Education Conference*. 10.1109/FIE.2011.6142913

Locke, E. A., & Latham, G. P. (1990). *A theory of goal setting and task performance*. Englewood Cliffs, NJ: Prentice Hall.

Lugmayr, A., Sutinen, E., Suhonen, J., Sedano, C. I., Hlavacs, H., & Montero, C. S. (2017). Serious storytelling – a first definition and review. *Multimedia Tools and Applications*, *76*(14), 15707–15733. doi:10.100711042-016-3865-5

Mackenzie, D. L. (2006). *What Works in Corrections*. Cambridge: Cambridge University Press. doi:10.1017/CBO9780511499470

Madden, M., Lenhart, A., Maeve, D., Sandra, C., & Gasser, U. (2013). *Teens and Technology 2013*. Washington, DC: Pew Research Center.

Mahayuddin, Z.R., Suwadi, N.A., Jenal, R., & Arshad, H. (2018). T. Implementing smart mobile application to achieve a sustainable campus. *International Journal of Supply Chain Management*, *7*(3), 154–159.

Malliarakis, C., Tomos, F., Shabalina, O., Mozelius, P., & Balan, O. (2015). How to Build an Ineffective Serious Game: Worst Practices in Serious Game Design. *ECGBL*.

Malone, T. W., & Lepper, M. R. (1987). Making learning fun: A taxonomy of intrinsic motivations for learning. In R. E. Snow & M. J. Farr (Eds.), Aptitude, learning, and instruction: Vol. 3. Conative and affective process analyses (pp. 223-253). Hillsdale, NJ: Lawrence Erlbaum.

Maloney, E. (2007). What Web 2.0 can teach us about learning. *The Chronicle of Higher Education*, *25*(18), B26.

Markopoulos, A. P., Fragkou, A., Kasidiaris, P. D., & Davim, J. P. (2015). Gamification in engineering education and professional training. *International Journal of Mechanical Engineering Education*, *43*(2), 118–131. doi:10.1177/0306419015591324

Marquez, J., Villanueva, J., Solarte, Z., & Garcia, A. (2016). IoT in education: Integration of objects with virtual academic communities. *New Advances in Information Systems and Technologies*, *1*, 201-212. Available: https://link.springer.com/chapter/10.1007/978-3-319-31232-3_19

Martin, J. (2017). Editorial: Learning Strategies in Engineering Education Using Virtual and Augmented Reality Technologies. *EURASIA Journal of Mathematics Science and Technology Education*, *13*(2), 297–300. Available: http://www.ejmste.com

McWherter, J., & Gowell, S. (2012). *Professional mobile application development*. Indianapolis, IN: John Wiley & Sons, Inc.

Mekler, E. D., Brühlmann, F., Tuch, A. N., & Opwis, K. (2015). Towards understanding the effects of individual gamification elements on intrinsic motivation and performance. *Computers in Human Behavior*. doi:10.1016/j.chb.2015.08.048

Membrillo-Hernández, J., Ramírez-Cadena, M. de J., Caballero-Valdés, C., Ganem-Corvera, R., Bustamante-Bello, R., Ordóñez-Díaz, J. A. B., & Elizalde, H. (2018). Challenge-Based Learning: The case of sustainable development engineering at the Tecnológico de Monterrey, Mexico City campus. *International Journal of Engineering Pedagogy*, *8*(3), 137–144. doi:10.3991/ijep.v8i3.8007

Minch, C. (2018). How to Write a Literature Review. *Centre for Effective Services*. Available: https://www.teachingcouncil.ie/en/_fileupload/Research/Literature-Review-Webinar.pdf

Ministero dello Svilupo Economico. (2019). *Piano Nazionale Impresa 4.0*. Retrieved from https://www.mise.gov.it/index.php/it/industria40

Mishra, A. S., Jha, J. K., & Umre, S. K. (2017). Mobile app and the library services. *International Journal of Information Library and Society*, *6*(2), 27–32.

Moncef Nehdi, P. E. (2002). Crisis of civil engineering education in information technology age: Analysis and prospects. *Journal of Professional Issues in Engineering Education and Practice*, *128*(3), 131–137. doi:10.1061/(ASCE)1052-3928(2002)128:3(131)

Morales-Menendez, R., Cantú Ortiz, F., Galeano Ramirez, N., Fangmeyer, J., & Hernández de Menéndez, A. M. (2019). *2019 IEEE Global Engineering Education Conference (EDUCON)*. Dubai, UAE: IEEE.

Müller, B. C., Reise, C., & Seliger, G. (2015). Gamification in factory management education. *Procedia CIRP*, *26*, 121–126. doi:10.1016/j.procir.2014.07.056

National Academy of Engineering. (2004). *The engineer of 2020*. Washington, DC: National Academies Press.

Nelson, L. M. (1998). *Collaborative problem solving: An instructional theory for learning through small group interaction* (Unpublished doctoral dissertation). Indiana University.

Nelson, L. M. (1999). Collaborative problem solving. In C. M. Reigeluth (Ed.), *Instructional-Design Theories and Models* (Vol. 2, pp. 241–269). Lawrence Erlbaum Associates.

Nickerson, J. V., Corter, J. E., Esche, S. K., & Chassapis, C. (2007). A model for evaluating the effectiveness of remote engineering laboratories and simulations in education. *Computers & Education*, *49*(3), 708–725. doi:10.1016/j.compedu.2005.11.019

NMC- The New Media Consortium. (2016). *NMC Horizon Report (2016): Higher Education Edition*. Austin, TX: The New Media Consortium. Available: https://www.sconul.ac.uk/sites/default/files/documents/2016-nmc-horizon-report-he-EN-1.pdf

Nordrum, A. (2016). *Popular Internet of Things Forecast of 50 Billion Devices by 2020 Is Outdated*. Available: https://spectrum.ieee.org

Norman, G. R., & Schmidt, H. G. (1992). The psychological basis of problem-based learning: A review of the evidence. *Academic Medicine*, *67*(9), 557–565. doi:10.1097/00001888-199209000-00002 PMID:1520409

O'Reilly, T. (2005). *What is web 2.0: Design patterns and business models for the next generation of software.* Retrieved 9 from http://oreilly.com/web2/archive/what-is-web-20.html

Ohland, M. W., Sheppard, S. D., Lichtenstein, G., Eris, O., Chachra, D., & Layton, R. A. (2008). Persistence, engagement, and migration in engineering programs. *Journal of Engineering Education*, *97*(3), 259–278. doi:10.1002/j.2168-9830.2008.tb00978.x

Ojasalo, J. (2001). Managing customer expectations in professional services. *Managing Service Quality*, *11*(3), 200–212. doi:10.1108/09604520110391379

Oliva, R., & Watson, N. (2010). Cross-functional alignment in supply chain planning: A case study of sales and operations planning. *Journal of Operations Management*, *29*(5), 434–448. doi:10.1016/j.jom.2010.11.012

Orlanda, Ramb, Langc, Houserd, Klinge, & Coccia. (2014). Saving energy in an office environment: A serious game intervention. *Energy and Buildings, 74,* 43-52. doi:10.1016/j.enbuild.2014.01.036

Ossiannilsson, E. (2018). Promoting active and meaningful learning for digital learners. In Handbook of Research on Mobile Technology, Constructivism, and Meaningful Learning (pp. 294-315). IGI Global. doi:10.4018/978-1-5225-3949-0.ch016

Özbay, Ö. (2015). Data Mining Concept and Data Mining Applications in Education. *International Journal of Educational Sciences*, *2*(5), 262–272.

Özbay, Ö. (2015). The concept of data mining and data mining applications in education. *International Journal of Educational Sciences*, (5): 262–272.

Özdemir, Ş. (2016). *An application on data mining in education and predicting student academic achievement* (PhD Thesis). Istanbul University, Institute of Science and Technology, İstanbul, Turkey.

Paoli, C. (2012). Google, Microsoft, & Apple Agree on Mobile Privacy Accord. *ADTMag*. https://adtmag.com/articles/2012/02/23/mobile-privacy-accord.aspx

Parasuraman, A., Zeithaml, V. A., & Berry, L. L. (1991). SERVQUAL: A Multiple-Item Scale for Measuring Consumer Perceptions of Service Quality. *Journal of Retailing*, *64*(1), 12–40.

Park, M., & Wang, L. (2019). Instructional design of technology-enhanced process writing for secondary EFL learners in Hong Kong. In J.-B. Son (Ed.), Context-Specific Computer-Assisted Language Learning (pp. 122-136). Queensland, Toowoomba, Australia: Asia-Pacific Association for Computer-Assisted Language Learning (APACALL).

Pasin, F., & Giroux, H. (2011). The impact of a simulation game on operations management education. *Computers & Education*, *57*(1), 1240–1254. doi:10.1016/j.compedu.2010.12.006

Patel, P., & Cassou, D. (2015). Enabling high-level application development for the Internet of Things. *Journal of Systems and Software, 103*, 62–84. doi:10.1016/j.jss.2015.01.027

Peat, M., Taylor, C. E., & Franklin, S. (2005). Re-engineering of undergraduate science curricula to emphasise development of lifelong learning skills. *Innovations in Education and Teaching International, 42*(2), 135–146. doi:10.1080/14703290500062482

Pereira, C. (2014). *Perspectives and Approaches for the Internet of Things* (Master Thesis). Faculty of Science and Technology, NOVA University Lisbon.

Pereira, J. (2018). *Implantação de módulos educacionais para circuitos elétricos e eletrônicos em universidades brasileiras no âmbito do projeto VISIR+*. Retrieved from https://repositorio.ufsc.br/handle/123456789/189331

Pereira, J., Silva, I. N., Simão, J. P., Carlos, L. M., Silva, J. B., & Bilessimo, S. M. S. (2017). *Modelo de repositório de práticas didáticas de circuitos elétricos e eletrônicos usando o laboratório remoto VISIR*. Presented at the Congresso Brasileiro de Educação em Engenharia - COBENGE.

Petersen, K., Feldt, R., Mujtaba, S., & Mattsson, M. (2008). Systematic Mapping Studies in Software Engineering. *EASE, 8*, 68–77.

Peterson, C. (2003). Bringing ADDIE to life: Instructional Design at Its Best. *Journal of Educational Multimedia and Hypermedia, 12*(3), 227–241.

Peters, T. J., & Waterman, R. H. (1982). *In Search of Excellence: Lessons from America's Best-run Companies*. New York, NY: Harper & Row.

Polat, S. (2006). *Kurumsal Kültür Organizasyonlar Kurallar Kurumlar*. SPK.

Prasad, J., Goswami, A., Kumbhani, B., Mishra, C., Tyagi, H., Jun, J. H., ... Das, S. K. (2018). Engineering curriculum development based on education theories. *Current Science, 114*(9), 1829. doi:10.18520/cs/v114/i09/1829-1834

ProMexico. (2017). *Crafting the Future. A roadmap for Industry 4.0 in Mexico*. Retrieved from https://clusterinstitute.com/Documentos/MRT_Industry_I40.pdf

Przybylski, A. K., Rigby, C. S., & Ryan, R. M. (2010). A motivational model of video game engagement. *Review of General Psychology, 14*(2), 154–166. doi:10.1037/a0019440

Qian, M., & Clark, K. (2016). Game-based Learning and 21st century skills: A review of recent research. *Computers in Human Behavior, 63*, 50–58. doi:10.1016/j.chb.2016.05.023

Raikar, M., Desai, P., Vijayalakshmi, M., & Narayankar, P. (2018). *Upsurge of IoT (Internet of Things) in engineering education: A case study*. Available: https://ieeexplore.ieee.org/document/8554546/

Rainer-Granados, J.J. (2018). *Informe LIBS Teaching Innovation Fund 2017/18. The UoL 4.0 initiative*. Internal communication.

Raju, P. K. (2003). Educating engineers for the information age. In *Proceedings of American Society for Engineering Education Annual Conference & Exposition*, (Session 2558, pp.8459.1-8459.14). Academic Press.

Ramirez-Mendoza, R. A., Cruz-Matus, L. A., Vazquez-Lepe, E., Rios, H., Cabeza-Azpiazu, L., Siller, H., & Ahuett-Garza, H. (2018). Towards a Disruptive Active Learning Engineering Education. *2018 IEEE Global Engineering Education Conference (EDUCON)*. 10.1109/EDUCON.2018.8363373

Ravitz, J. (2009). Introduction: Summarizing Findings and Looking Ahead to a New Generation of PBL Research. *Interdisciplinary Journal of Problem-Based Learning, 3*(1).

Reber, A. S. (1989). Implicit learning and tacit knowledge. *Journal of Experimental Psychology. General, 118*(3), 219–235. doi:10.1037/0096-3445.118.3.219 PMID:2527948

Reddy. (2018). *Engineering Education in India – Short & Medium Term Perspectives*. Available https://www.aicte-india.org

Reigeluth, C. M., Watson, W. R., Watson, S. L., Dutta, P., Chen, Z., & Powell, N. D. P. (2008). Roles for technology in the information-age paradigm of education: Learning management systems. *Educational Technology, 48*(6), 32–39.

Richardson, W. (2009). *Blogs, wikis, podcasts, and other powerful web tools for classrooms* (2nd ed.). Thousand Oaks, CA: Corwin Press.

Richter, A. (2015). LEARNING 4.0: Virtual Immersive Engineering Education. *International Best Practices and Applications, 11*, 51–66.

Rieber, L. P. (1996). Seriously considering play: Designing interactive learning environments based on the blending of microworlds, simulations, and games. *Educational Technology Research and Development, 44*(2), 43–58. doi:10.1007/BF02300540

Rigby, C. S., & Ryan, R. M. (2011). *Glued to games: How video games draw us in and hold us spellbound*. Santa Barbara: Praeger.

Ringstaff, C., & Yocam, K. (1996). *Integrating Technology into Classroom Instruction: An assessment of the impact of the ACOT teacher development center project*. Apple Classrooms of Tomorrow. ACOT report #22. Retrieved from https://www.apple.com/euro/pdfs/acotlibrary/rpt22.pdf

Riojas, M., Lysecky, S., & Rozenblit, J. (2012). Educational Technologies for Precollege Engineering Education. *IEEE Transactions on Learning Technologies, 5*(1), 20–37. doi:10.1109/TLT.2011.16

Rodrigues, A. L. (2016). A integração pedagógica das tecnologias digitais na foração ativa de professores. Presented at the *IV Congresso Internacional das TIC na Educação: Tecnologias digitais e a Escola do Futuro*.

Rokeach, M. (1983). *The nature of human values*. New York: The Free Press.

Romero, C., & Ventura, S. (2007). "Educational Data Mining: A Survey from 1995 to2005". Expert Systems with Applications,Cilt: 33. *Sayı, 1*, 135–146.

Rondon, S., Sassi, F. C., & Furquim de Andrade, C. R. (2013). Computer game-based and traditional learning method: A comparison regarding students' knowledge retention. *BMC Medical Education, 13*(1), 30. doi:10.1186/1472-6920-13-30 PMID:23442203

Roque, G. R., Silva, I. N., Alves, G. R., Alves, J. B., Silva, J. B., & Bilessimo, S. M. S. (2017). *Utilização do laboratório remoto VISIR como recurso educacional num curso de engenharia mecatrônica.* Presented at the Congresso Brasileiro de Educação em Engenharia - COBENGE.

Rosebrough, T. M., & Leverett, R. G. (2011). *Transformational teaching in the information age.* Alexandria, VA: ASCD Publication.

Rosen, M. (2009). Engineering Education: Future Trends and Advances. *Proceedings of the 6th WSEAS International Conference on Engineering Education.* Available: https://pdfs.semanticscholar.org/c85b/7aa6297ce266065bceed4d46b4851252bd31.pdf

Rotgans, J. I., & Schmidt, H. G. (2011). Cognitive engagement in the problem-based learning classroom. *Advances in Health Sciences Education: Theory and Practice, 16*(4), 465–479. doi:10.100710459-011-9272-9 PMID:21243425

Rucker, M. (2016). *A Brief Overview of Three Types of Literature Review.* Available: https://unstick.me/brief-overview-three-types-literature-review/

Ruth, G. (2017). *The global state-of-the-art in engineering education Outcomes of Phase 1 benchmarking study.* Available: https://www.cti-commission.fr/wp-content/uploads/2017/10/Phase-1-engineering-education-benchmarking-study-2017.pdf

Ruth, G. (2018). *The global state of the art in engineering education.* School of Engineering Massachusetts Institute of Technology. Available: http://neet.mit.edu/wp-content/uploads/(2018)/03/MIT_NEET_GlobalStateEngineeringEducation(2018).pdf

Ryan, R. M., Rigby, C. S., & Przybylski, A. K. (2006). The motivational pull of video games: A self-determination theory approach. *Motivation and Emotion, 30*(4), 344–360. doi:10.100711031-006-9051-8

Saaty, T. L. (1990). How to make a decision: The Analytic Hierarchy Process. *European Journal of Operational Research, 48*(1), 9–26. doi:10.1016/0377-2217(90)90057-I

Saaty, T. L., & Vargas, L. G. (2001). *Models, Methods, Concepts and Applications of the Analytic Hierarchy Process.* Norwell: Kluwer Academic Publishers. doi:10.1007/978-1-4615-1665-1

Sailer, M., Hense, J. U., Mayr, S. K., & Mandl, H. (2017). Jan Ulrich Hense, Sarah Katharina Mayr, Heinz Mandl, 2017, How gamification motivates: An experimental study of the effects of specific game design elements on psychological need satisfaction. *Computers in Human Behavior, 69*, 371–380. doi:10.1016/j.chb.2016.12.033

Salaheddin, O. (2014). *Experiências da Aplicação de VISIR na Universidade de Al-Quds*. Presented at the *11th International Conference On Remote Engineering And Virtual Instrumentation*, Portugal.

Sánchez, J., & Olivares, R. (2011). Problem solving and cooperation using mobile serious games. *Computers & Education*, *57*(3), 1943–1952. doi:10.1016/j.compedu.2011.04.012

Schein, E. (1992). *Organizational Culture and Leadership*. New York, NY: Random House.

Schippers, M., Rook, L., & van de Velde, S. (2012). Team supply chain management decisions: Curvilinear effects of reflexitivity and regulatory focus. Rotterdam: Academic Press.

Schippers, M. C., Homan, A. C., & Knippenberg, D. V. A. N. (2012, February). To reflect or not to reflect: Prior team performance as a boundary condition of the effects of reflexivity on learning and final team performance †. *Journal of Organizational Behavior*, *34*(1), 6–23. doi:10.1002/job.1784

Schuster, K., Grob, K., Vossen, R., Richert, A., & Jeschke, S. (2017). Preparing for Industry 4.0 – Collaborative Virtual Learning Environments in Engineering Education. In *Engineering Education 4.0*. Cham: Springer.

Schuster, K., Plumanns, L., Groß, K., Vossen, R., Richert, A., & Jeschke, S. (2015). Preparing for Industry 4.0 – Testing Collaborative Virtual Learning Environments with Students and Professional Trainers. *International Journal of Advanced Corporate Learning*, *8*(4), 14. doi:10.3991/ijac.v8i4.4911

Şendağ, S., & Odabaşı, H. F. (2009). Effects of an online problem based learning course on content knowledge acquisition and critical thinking skills. *Computers & Education*, *53*(1), 132–141. doi:10.1016/j.compedu.2009.01.008

Şengür, D., & Tekin, A. (2013). Estimation of Graduation Grades of Students with Data Mining Methods. *Journal of Information Technologies, 6*(3).

Serdarasan, S. (2013). A review of supply chain complexity. drivers. *Computers & Industrial Engineering*, *66*(3), 533–540. doi:10.1016/j.cie.2012.12.008

Sharma, R. R. (2017). A Competency Model for Management Education for Sustainability, 2017. *Vision (Basel)*, *21*(2), 10–15.

Shi, L., Cristea, A. I., Hadzidedic, S., & Dervishalidovic, N. (2014). Contextual gamification of social interaction towards increasing motivation in social E-learning. In *Advances in web-based learning-ICWL 2014* (pp. 116–122). Springer. doi:10.1007/978-3-319-09635-3_12

Silva, I. N. (2016). *Uma proposta de aplicação de experimentação remota e dispositivos móveis em aulas experimentais de física no ensino médio*. Presented at the SICT-SUL.

Silva, I. N., Pereira, J., Bilessimo, S., Alves, G., & Silva, J. B. (2018). *O Laboratório VISIR: uma revisão sistemática*. Presented at the Simpósio Íbero-Americano de Tecnologias Educacionais.

Silva, I. N., Roque, G. R., Bilessimo, S. M. S., Alves, G. R., & Silva, J. B. (2018). *Laboratório Remoto VISIR: estudo de caso da integração na plataforma RELLE*. Retrieved from https://www.aacademica.org/congreso.aahd2018/tabs/abstracts

Silva, K. C. N. (2018). *Inovação social na educação básica: um estudo de caso sobre o Laboratório de Experimentação Remota da Universidade Federal de Santa Catarina*. Retrieved from https://repositorio.ufsc.br/handle/123456789/191134

Silva, K. C. N., Silva, I. N., & Bilessimo, S. M. S. (2017). *Oficinas de eletrônica básica para estudantes do ensino médio de uma escola pública de educação básica*. Presented at the Simpósio Íbero-Americano de Tecnologias Educacionais.

Silva, P. F., & Melo, S. D. G. (2018). O trabalho docente nos Institutos Federais no contexto de expansão da educação superior. *Educação e Pesquisa*, *44*(0). doi:10.15901678-4634201844177066

Silveira, M. S., & Carneiro, M. L. F. (2012). *Diretrizes para a Avaliação da Usabilidade de Objetos de Aprendizagem*. SBIE.

Simão, J. P. S. (2018). *Modelo para registro de dados de experiência de aprendizagem em laboratórios remotos*. Retrieved from https://repositorio.ufsc.br/handle/123456789/191118

Smith, S., & Chan, S. (2017). Collaborative and Competitive Video Games for Teaching Computing in Higher Education. *Journal of Science Education and Technology*, *26*(4), 438–457. doi:10.100710956-017-9690-4

Subhash, S., & Cudney, E. (2018). Gamified learning in higher education: A systematic review of the literature. *Computers in Human Behavior*, *87*, 192–206. doi:10.1016/j.chb.2018.05.028

Suda, L. V. (2016). *NASA's Project Management Leadership LAB*. Academic Press.

Sullivan, L. P. (1986). Quality function deployment. *Quality Progress*, *19*, 39–50.

Šundov, J., & Dulčić, Z. (2017). Managing by values – Top managers impact on the process of creating new values for the company. *5th International OFEL Conference on Governance, Management, and Entrepreneurship The Paradoxes of Leadership and Governance in the Postmodern Society*.

Sutherland, R., Robertson, S., & John, P. (2004). Interactive education: Teaching and learning in the information age. *Journal of Computer Assisted Learning*, *20*(6), 410–412. doi:10.1111/j.1365-2729.2004.00100.x

Tawfik, M. (2013). Virtual instrument systems in reality (visir) for remote wiring and measurement of electronic circuits on breadboard. *IEEE Transactions on Learning Technologies*. 10.1109/TLT.2012.20

Thomas, E. H., & Galambos, N. (2004). What satisfies students? Mining studentopinion data with regression and decision tree analysis. *Research in Higher Education*, *45*(3), 251–269. doi:10.1023/B:RIHE.0000019589.79439.6e

Thomé, A. M. T., Scavarda, L. F., Fernandez, N. S., & Scavarda, A. J. (2012). Sales and operations planning and the firm performance. *International Journal of Productivity and Performance Management, 61*(4), 359–381. doi:10.1108/17410401211212643

Tinto, V. (1993). *Leaving college: Rethinking the causes and cures of student attrition.* Chicago, IL: The University of Chicago Press.

TMS. (2019). *University Accreditation ABET.* Retrieved from https://www.tms.org/portal/PROFESSIONAL_DEVELOPMENT/Professional_Development_Resources/University_Accreditation__ABET_/portal/Professional_Development/Professional_Development_Resources/University_Accreditation__ABET_.aspx?hkey=6efc343b-effd-40d3-8cb2-b5c3934981ef

Topalli, D., & Cagiltay, N. E. (2018). Improving programming skills in engineering education through problem-based game projects with Scratch. *Computers & Education, 120*, 64–74. doi:10.1016/j.compedu.2018.01.011

Töremen, F., & Akdemir, D. A. (2007). Eğitim örgütlerinin kişiliğine uygun paradigma: Değerlerle yönetim. *E-Journal of New World Sciences Academy*, *2*(4), 1306–1311.

Triki, N. (2016). Innovations and Curriculum Development for Technical and Engineering Education in Libya. *Literacy Information and Computer Education Journal*, *7*(4). Available https://infonomics-society.org/licej/. doi:10.20533/licej.2040.2589.2016.0324

Tufekci, A., & Guzeldereli, E. A. (2018). Educational Data Mining In Distance education: A Sysyematic Literature Mapping Study. *International Women Online Journal of Distance Education, 7*(3), 61–73.

University of Lincoln. (2016). *Thinking Ahead 2016 -2021. University of Lincoln. Strategic Plan.* Retrieved from https://www.lincoln.ac.uk/home/media/responsive2017/abouttheuniversity/managementandstrategy/UOL,Strategic,Plan,(MAR,2016),V5Final.pdf.

University of Lincoln. (2019). *Student as Producer.* Retrieved from https://studentasproducer.lincoln.ac.uk/

Vaidya, S., & Kumar, S. (2006). Analytic hierarchy process: An overview of applications. *European Journal of Operational Research, 169*(1), 1–29. doi:10.1016/j.ejor.2004.04.028

Valoris, M. (2015). A mixed-methods analysiis of best practices for land-grant unıversity mobile applications from a user experience design perspective (Master's thesis). Colorado State University. Libraries.

Veletsianos, G. (2016). *Emergence and innovation in digital learning: Foundations and applications.* Athabasca University Press. doi:10.15215/aupress/9781771991490.01

Villari, M., Al-Anbuky, A., Celesti, A., & Moessner, M. (2016). Leveraging the Internet of Things: Integration of Sensors and Cloud Computing Systems. *SAGE Journal.* Available on https://journals.sagepub.com/doi/full/10.1177/155014779764287

Vujović, V., Maksimović, M., & Perišić, B. (2014). Collaboration in Software Engineering classroom. *12th IEEE International Conference on Emerging eLearning Technologies and Applications*, 40(6), 505-510.

Vurgun, L., & Öztop, S. (2011). Yönetim ve örgüt kültüründe değerlerin önemi. *Süleyman Demirel Üniversitesi İktisadi ve İdari Bilimler Fakültesi Dergisi*, *16*(3), 217–230.

Vygotsky, L. S. (1980). *Mind in society: The development of higher psychological processes*. Harvard University Press. doi:10.2307/j.ctvjf9vz4

Waddoups, G. L., Wentworth, N., & Earle, R. (2004). Principles of technology integration and curriculum development. *Computers in the Schools*, *21*(1-2), 15–23. doi:10.1300/J025v21n01_02

Walters, D. (2002). *Value and value chains in educations. Operations Strategy*. Red Globe Press.

Wang, Y. (2002). From teacher-centredness to student-centredness: Are preservice teachers making the conceptual shift when teaching in information age classrooms? *Educational Media International*, *39*(3-4), 257–265. doi:10.1080/09523980210166710

Watson, W. E., BarNir, A., & Pavur, R. (2005). Cultural diversity and learning teams: The impact on desired academic team processes. *International Journal of Intercultural Relations*, *29*(4), 449–467. doi:10.1016/j.ijintrel.2005.06.001

Watson, W. E., Johnos, L., Kumar, K., & Critelli, J. (1998). Process gain and process loss: Comparing interpersonal processes and performance of culturally diverse and non-diverse teams across time. *International Journal of Intercultural Relations*, *22*(4), 409–430. doi:10.1016/S0147-1767(98)00016-9

Weenk, E. (2019). *Mastering the Supply Chain: Principles, Practice and Real-Life Applications*. Kogan Page Publishers.

WEF- World Economic Forum. (2018). *The Future of Jobs Report (2018), Insight Report*. Available: http://www3.weforum.org/docs/WEF_Future_of_Jobs_(2018).pdf

WEF. (2018). The Future of Jobs Report 2018. *World Economic Forum*. Retrieved January 6, 2020 from http://www3.weforum.org/docs/WEF_Future_of_Jobs_2018.pdf

WEF-World Economic Forum. (2015). *Deep Shift Technology Tipping Points and Societal Impact: Survey Report, September 2015*. Available: http://www3.weforum.org/docs/WEF_GAC15_Technological_Tipping_Points_report_2015.pdf

Wenger, E. (2010). *Communities of Practice and Social Learning Systems: the Career of a Concept*. Social Learning Systems And Communities Of Practice.

Werbach, K., & Hunter, D. (2012). *For the win: How game thinking can revolutionize your business*. Philadelphia: Wharton Digital Press.

Wertz, P. (2016). *Engineer receives funding to support teaching in a virtual environment*. Available: https://news.psu.edu/

West, M. A. (2012). Creating Effective Teams. In Effective Teamwork: Practical Lessons from Organizational Research (3rd ed., pp. 1–12). John Wiley & Sons, Ltd. and the British Psychological Society.

Wiggins, B. E. (2016). An overview and study on the use of games, simulations, and gamification in higher education. *International Journal of Game-Based Learning*, *6*(1), 18–29. doi:10.4018/IJGBL.2016010102

Wilkinson, P. (2016). A Brief History of Serious Games. Entertainment Computing and Serious Games Lecture Notes in Computer Science, 17-41. doi:10.1007/978-3-319-46152-6_2

World Bank. (2019). *The Changing Nature of Work*. Retrieved from http://documents.worldbank.org/curated/en/816281518818814423/pdf/2019-WDR-Report.pdf

World Economic Forum. (2016). The Future of Jobs Employment, Skills and Workforce Strategy for the Fourth Industrial Revolution. *Global Challenge Insight Report*. Available: http://www3.weforum.org/docs/WEF_Future_of_Jobs.pdf

World Economic Forum. (2018). *The future of jobs*. Retrieved from http://www3.weforum.org/docs/WEF_Future_of_Jobs_2018.pdf

Wouters, P., Spek, E. D., & Oostendorp, H. (2011). Measuring learning in serious games: A case study with structural assessment. *Educational Technology Research and Development*, *59*(6), 741–763. doi:10.100711423-010-9183-0

Wu, W.-H., Hsiao, H.-C., Wu, P.-L., Lin, C.-H., & Huang, S.-H. (2012). Investigating the learning-theory foundations of game-based learning: A meta-analysis. *Journal of Computer Assisted Learning*, *28*(3), 265–279. doi:10.1111/j.1365-2729.2011.00437.x

Yee, N. (2006). Motivations for play in online games. *Cyberpsychology & Behavior*, *9*(6), 772–777. doi:10.1089/cpb.2006.9.772 PMID:17201605

Yew, E. H., & Goh, K. (2016). Problem-based learning: An overview of its process and impact on learning. *Health Profession Education*, *2*(2), 75–79. doi:10.1016/j.hpe.2016.01.004

Yıldız, S., & Kurtuldu, H. S. (2013). Factors Affecting Electronic Service Brand Equity. In *Transcultural Marketing for Incremental and Radical Innovation* (pp. 434–492). IGI Global.

Zeithaml, V. A., Berry, L. L., & Parasuraman, A. (1993). The Nature and Determinants of Customer Expectations of Service. *Journal of the Academy of Marketing Science*, *21*(1), 1–1. doi:10.1177/0092070393211001

Zhou, J., Leppanen, T., & Harjula, E. (2013). Cloud Thing: A Common Architecture for Integrating the Internet of thing with Cloud Computing. In *CSCWD, 2013*. IEEE.

Zhuang, T., & Xu, X. (2018). New engineering education in Chinese higher education: Prospects and challenges. *Tuning Journal for Higher Education*, *6*(1), 69–109. doi:10.18543/tjhe-6(1)-2018pp69-109

Zhu, W., Marquez, A., & Yoo, J. (2015). Engineering economics jeopardy! Mobile app for university students. *The Engineering Economist*, *60*(4), 291–306. doi:10.1080/0013791X.2015.1067343

Zichermann, G., & Cunningham, C. (2011). Gamification by design: Implementing game machanics in web and mobile apps. Cambridge.

About the Contributors

Şeyda SerdarAsan is an Assistant Professor of Industrial Engineering at the Istanbul Technical University since 2012. She received her BSc and MSc degrees from Istanbul Technical University in 1999 and 2001, and her Doctoral degree from Technische Universität Berlin in 2009. Her current research interests focus on excellence in engineering education and digital supply chain management.

Erkan Isikli is an assistant professor in the Industrial Engineering Department at Istanbul Technical University (ITU). He has also been serving as ABET program coordinator at the same department since 2015. Having a Ph.D from the Industrial and Systems Engineering Department of Wayne State University and a B.Sc. in Mathematics Engineering from ITU, Dr. Isikli has specialized in constructing statistical models for real-world problems in various fields such as education, environment, healthcare, marketing, non-profit organizations, sports, and supply chain. He has published several journal papers and book chapters and presented many proceedings at esteemed conferences worldwide.

* * *

Ali Sharaf Al-Musawi has obtained his PhD on learning resources and technology centers in 1995 from Southampton University, UK. He works for the Sultan Qaboos University since 1985. At present, he is a professor at the Instructional and Learning Technologies Department at the College of Education. He has published numerous journal research articles, chapters in reviewed books, and papers; and contributed in many conferences, symposia, and workshops. He conducted and published several national, regional, and Arab studies and reports. He wrote a book on cooperative learning in 1992, contributed in writing another in 2003 and an educational lexicon in 2014; and published a book on learning resources and technology centers in 2004. He also translated, with others, two books on e-learning strategies and instructional multimedia to Arabic in 2005 and 2010. Prof. Ali has several activities in the fields of instructional skills development, study skills, instructional design, and web-based design; his interests include Arabic poetry; he published two anthologies in addition to other hand-written ones.

Issa Alghatrifi is a lecturer at the IT studies Dept, Nizwa College of Technology in the Omani Ministry of Manpower. He worked as a supervisor of Omani private schools at MOE after obtaining his MSc from the UK. He is currently pursuing his PhD studies in Malaysia in the field of IoT engineering.

Özge Andiç-Çakir, an Associate Professor at the Department of Civil Engineering of Ege University, has been worked as Vice Director (2013-2016) and board member (2016-2018) of Ege University Science and Technology Centre-Technology Transfer Office (EBILTEM-TTO) which is a regional and national expertise organization on innovation management, R&D and technology transfer. Currently, she is the Vice Director of Ege University Application and Research Center for Testing and Analysis (EGE-MATAL). She got her Bachelor's, Master's and Ph.D. degrees at Ege University and worked at TU Delft, Netherlands as Turkish Council of Higher Education research fellow following her studies. She has participated in several collaborative projects as well as being project manager. She has managed nEBULA project funded under ERASMUS + programme by Turkish National Agency and coordinated ee-WiSE project on behalf of Ege University funded under EU FP7 which is related with energy efficient retrofitting. Currently, she is the coordinator of Engineering Student Centered Learning Approaches (ESCOLA) project funded under ERASMUS + programme by Turkish National Agency. She also has valuable experience in organization and development of dissemination activities like conferences, seminars, workshops, etc. Dr. ANDİÇ-ÇAKIR has been teaching on concrete technologies, testing and analysis techniques and characterization of engineering materials. Until now, she has actively involved in 31 projects and published 77 papers and prepared 7 book chapters/reports.

Cahit Ali Bayraktar is an academic of Industrial Engineering at Istanbul Technical University. He graduated with a BS from Istanbul Technical University in 1990, a MS from Istanbul Technical University in 1993 and a PhD from Istanbul Technical University in 2001. His current research interests include Management and Organization, Strategic Management, Behavioral Sciences and Human Resources Management.

Simone Bilessimo graduated in Mechanical Production Engineering from the Federal University of Santa Catarina (1997), Master's degree (1999) and PhD (2007) in Production Engineering from the Federal University of Santa Catarina. She is currently a dedicated teacher (Assistant Professor IV) at the Federal University of

Santa Catarina - Campus Araranguá, including as a teacher in the Postgraduate Program in Information and Communication Technologies. He has been active in the following subjects: entrepreneurship, innovation management, integration of technology in education, as well as the use of Information and Communication Technologies in education and digital business. Experience with event organization. Participates in the Research Group RExLab - Remote Experimentation Laboratory.

Hasan Çakır is an Associate Professor in the department of Computer Education and Instructional Technology at Gazi University, Ankara- Turkey. Dr. Çakır received his PhD degree from the Instructional Systems Technology Department of Indiana University-Bloomington under the guidance of Barbara A. Bichelmeyer and Thomas Duffy. His research focuses on mainly two areas; first one is supporting social constructivist teaching and learning methods with collaborative technologies and, second, creating design based learning environments supported with technology to engage students with meaningful learning activities. Dr. Çakır has published many national and international book chapters, articles, and project reports, has been teaching graduate and undergraduate level courses, and completed seven national and international projects on supporting teaching and learning with computer and information technologies.

Cemil Ceylan is an assistant professor in the Department of Industrial Engineering at Istanbul Technical University. He received a Ph.D. degree in Organizational Flexibility from Istanbul Technical University. He is currently working at Organizational Measurement, Process Management, Management, Organization, and Human Resources Management. He has several national and international publications on related topics.

Chang Ge is a Lecturer in People and Organisation Department, Lincoln International Business School, University of Lincoln. Through her role as the College Digital Lead, Chang is actively involved in supporting and mentoring faculty through the use of creative technologies; innovation in teaching and curriculum design. As a member of Digital Education Sub-Group (DESG) of University of Lincoln, Chang has led, designed and delivered some major training sessions in areas of digital communication to engage students and to develop their digital transferable skills. In Chartered Management Degree Apprenticeship (CMDA), Chang led a module on Digital Business and New Technologies in the Workplace, which promotes the development of a range of digital communication skills and knowledge that can be transferred to students' personal and professional context. Chang is in her second year of PhD in Education study at the University of Lincoln. Her research includes exploring the nature of what constitutes effective distance learning; pedagogical theories combined with technological approaches in relation to the design and delivery of online learning and teaching to engage distance learners; digital capabilities of students and academics in online learning and teaching environment.

Esra Ayca Guzeldereli Yilmaz graduated from Gazi University, Industrial Arts Education Faculty, and Computer Education Department in 2010, and completed her master's degree at Sakarya University, Graduate School of Natural And Applied Sciences, Department of Computer Education in 2012. In the same year, GUZELDERELI started to work as a lecturer at Afyon Kocatepe University, Emirdag Vocational School, Department of Computer Technologies. She worked as assistant manager for two years at the same institution. Since 2015, she has been doing a Ph.D. in Information Systems at Gazi University, Institute of Informatics. She has interests in distance education, data mining, artificial intelligence techniques, and human performance technology.

Meltem Karadayı has graduated from ITU Industrial Engineering Department in 2019 where she has successfully developed her own serious game as a graduation project. She currently works as a game analyst and level designer.

Ayşe Kılıç was born in Istanbul in 1985. She completed her secondary education at Kültür Fen Lisesi. She graduated from Istanbul Technical University as a Metallurgical and Materials Engineer in 2008. She has also taken the Master`s degree and doctorate degree from the same university department at the years respectively as 2010, 2018. Working currently as a Dr. Lecturer at ITU Metallurgical & Materials Engineering Department. On the other hand,, she is working as "Assessment & Measurement Specialist" in the Quality Coordination Unit. Every year, Higher Education Council of Turkey request a report from all universities about the evaluation of quality processes. Within this scope, survey designs are implemented for the general and service satisfaction of students, academic & administrative staff. The statistical evaluation of the survey results and reporting to the management is done periodically by a group of whom she is involved. Also she is an executive board member of ITU Center for Excellence in Education which was established to develop new approaches in line with the worldwide changes occurring in engineering education. Besides these her areas of Interest: Design and Material Selection, Engineering Education, Assessment and Evaluation Methods, Data Analysis, Engineering Ethics and Ethical Codes, New Engineering Product Development and Innovation, Engineering History, Non-Destructive Testing Methods (PT2, MT2, VT2, UT2, VT2), Quality and Accreditation in Engineering History, Statistical Process Control, Project Management.

Geeta Lakshmi is an Associate Professor in Finance with Lincoln Business School. She has been a Director of Study to three successful PhD completions, taught internationally and in the UK for several years and has mentored colleagues in areas of teaching development. She has been involved with research projects for the CDC, Department for International Development, UK and has untaken research funded by the UN. Currently, she serves as a Director, Sustainable Hockerton, having co-founded a village community company dealing in sustainable energy.

Rosario Michel-Villarreal is Associate Lecturer and PhD researcher at Lincoln International Business School. She holds a BSc in Industrial and Systems Engineering and a Masters in Engineering. She is a fellow of the Higher Education Academy and member of the Institution of Engineering and Technology (MIET), the Women's Engineering Society and the European Operations Management Association. Before joining academia, she worked in the food manufacturing industry as Quality Assurance Technician and Quality Engineer. Her research focus is on governance and its relationship to sustainability practices of short food supply chains. Since 2017, she has collaborated as research lead in the UoL4.0 Challenge project at the University of Lincoln. UoL4.0 Challenge project is based on the challenge-based learning approach and aims at the development of personal and professional abilities to better deal with the Digital world. It brings together students, businesses and government to tackle key challenges brought by Industry 4.0.

Fırat Sarsar is an Assistant Professor at Ege University, Department of Computer Education and Instructional Technology in Turkey. He earned his BA and MS degrees in the Department of Computer Education and Instructional Technology in Turkey. Dr. Sarsar graduated from Georgia State University in USA with his Ph.D. in Instructional Technology as a Fulbright scholar in 2014. His PhD was fully funded by Fulbright Scholarship. He taught "Computer Skills for the Information Age" course for undergraduate students at Georgia State University for 3 years. He was in many project related to technology integration and teacher education. He holds the Association for Educational Communications and Technology (AECT) Cochran Intern Award in 2013 which is given for recognition of young leaders of the field. Currently, He teaches "Social Media and Web2.0 tools in Education", "Internet for Educator", Technology Integration in Education" and "e-Learning Design" courses in undergrad and graduate levels in Ege University. His research interests are motivation, online learning, mobile technologies, e-learning and online feedback strategies. He is currently the Director of Distance Learning Application and Research Center and also Vice-Director of Ege University Continuing Education Center.

Isabela Silva is a Professor at SESI/SENAI Criciúma. Collaborator at the Remote Experimentation Laboratory (RExLab)/UFSC. Master in Information and Communication Technologies from the Federal University of Santa Catarina. Graduate in Information and Communication Technologies from the Federal University of Santa Catarina with a degree in Inclusive Education and Assistive Technology.

Juarez Bento da Silva has a degree in Business Administration from the Catholic University of Rio Grande do Sul (1991), specialization in Networking and Telecommunications at the UFSC(1999), Masters in Computer Science from UFSC (2002), PhD in Engineering and Knowledge Management, UFSC(2007). He is currently Associate Professor at UFSC and coordinator of the Remote Experimentation Lab (RExLab). He is currently Associate Professor at the Federal University of Santa Catarina (UFSC) and coordinator of the Remote Experimentation Lab (RExLab). He has experience in Computer Science with emphasis in Computer Systems and acting on the following topics: remote experimentation, 3D virtual worlds, embedded systems, computers in education and digital inclusion. The main interest in research and current focus is the development and use of new information and communication technologies in teaching and learning. Currently we seek to develop and implement new devices for remote testing and work on the integration of hardware (remote experiments) in 3D virtual worlds, using open-source project. This is the link to my resume for more details: http://lattes.cnpq.br/1594563006260546.

Ismail Yılmaz Taptik was born as a son of Dr. Nazmiye Taptik and Dr. Faruk Taptik in Ankara, 1954. He completed his secondary education at Istanbul Erkek Lisesi. He graduated from Istanbul Technical University as a Metallurgical Engineer in 1978. He also taken the Master`s degree (1979) and doctorate degree (1984) from the same university. During the year of 1986, he conducted scientific research about engineering economics and quality at University of Leoben (Austria) and also researches about non-destructive testing methods and quality issues at Berlin BAM in 1988. He became associate professor in 1990 and professor in 1996. He worked as the Head of ITU Metallurgical and Materials Engineering Department between the years 2007-2010 and as a founding dean of Students Affairs between the years 2010-2020. He is currently working as academic staff at ITU Chemical and Metallurgical Engineering Faculty and also as the founding director of ITU Center for Excellence in Education. Prof. Dr. Yilmaz Taptik has published around 120 articles and papers both in national and international journals and has books related to quality management. His area of interest are; non-destructive testing methods, engineering education, material selection and design, quality systems, project management, education and accreditation thermal barrier coating, etc. He is married and has a son.

Kutay Tinç, is a lecturer and researcher in Istanbul Technical University Industrial Engineering Department. He's a mathematical engineering graduate in the same university. He teaches courses like Theory of Probability, Operations Research and Game Theory. He also teaches in the M.Sc. program titled 'Game and Interaction Technologies'. He is interested in game design and game-based learning academically.

Aslıhan Tufekci is a lecturer at the Gazi University, Faculty of Education. Aslıhan Tufekci has completed the undergraduate studies at the Gazi University, Faculty of Industrial Arts and Education, at the section Computer Training in 1994. With completing the graduate study at the Gazi University, she started to work as a research assistant in the Faculty of Industrial Arts Education at Gazi University. In 1997, she completed her master's degree at Gazi University, Graduate School of Natural and Applied Sciences, and Department of Computer Education. Between the years of 1997-2002, she completed his Ph.D. education in Electronics and Computer Education Department of the same Institute. Since 2015, Aslıhan Tufekci has been working as an Associate Professor at Gazi University, Faculty of Education, Department of Computer and Instructional Technology Education and she is the director of Gazi University, Institute of Informatics. The interest of the author Tufekci is instructional technology and design, distance education and applications, artificial intelligence techniques and human performance technology.

Erhan Ünal is a research assistant PhD. in the department of Computer Education and Instructional Technology at Afyon Kocatepe University, Afyonkarahisar, Turkey. Dr. Ünal received his PhD degree from the Department of Computer Education and Instructional Technology at Gazi University. His research areas are designing constructivist learning environments with collaborative technologies, integrating web 2.0 technologies into education and technology adoption models. Dr. Ünal has published national and international book chapters and articles.

Eliseo Vilalta-Perdomo is an Associate Professor in Operations and Logistics Management, Lincoln International Business School, University of Lincoln. He has a professional experience both in the Government and Private Sectors. In the Office of the Speaker for the President of Mexico, he was the Documentation Deputy Director; in Aeropuertos y Servicios Auxiliares, Advisor for the Operations Deputy Director; in PEMEX Gas y Petroquímica Básica, General Superintendent of Natural Gas Sales, and in the Mexican Secretary of the Interior, Advisor to the Secretary. In the private sector he has been collaborating as Associate in Consultores y Promotores Asociados, S.C.; in Desarrolladora Metropolitana, S.A. de C.V. Assistant of the General Director, and in Grupo Tribasa, Evaluation Advisor for the International Vice-president. Previous academic experience includes part-time

Lecturer (1993-1999) and full-time (2000-2011). Associate Professor (Tecnologico de Monterrey), Head of the Department of Industrial & Systems Engineering (Tecnologico de Monterrey, Campus Guadalajara), and Dean of the Division of Higher Education (Tecnologico de Monterrey, Campus Irapuato). He has been awarded with a Doctorare Honoris Causa by Universidad Privada Antenor Orrego, Peru, and Fellowships of the Chartered Institute of Logistics and Transport, the Operational Research Society, and AdvanceHE (former Higher Education Academy).

Patrick Willems is an assistant professor in the Industrial Engineering Department at the Catholic University Leuven (KUL) and the Management and Technology Department of University College Leuven Limburg (UCLL). Patrick has a MSc. in Social and Military sciences, a MSc. in Management and an MBA. He has specialized in serious supply chain games in education. He is an education associate for Inchainge where he coaches professors from different countries, who play The Fresh Connection, The Cool Connection or more recently The Blue Connection. Next to these activities, he is a free-lance consultant for companies in the Retail business.

Index

H

I

L

M

O

P

R

S

T

U

V

www.ingramcontent.com/pod-product-compliance
Lightning Source LLC
LaVergne TN
LVHW061545290926
849162LV00004B/907